Wolves on the Hunt

Wolves on the Hunt

The Behavior of Wolves Hunting Wild Prey

L. DAVID MECH,

DOUGLAS W. SMITH, AND

DANIEL R. MACNULTY

With Supplementary Online Video by Robert K. Landis

THE UNIVERSITY OF CHICAGO PRESS   *Chicago and London*

L. David Mech is a senior research scientist with the US Geological Survey and an adjunct professor in the Department of Fisheries, Wildlife, and Conservation Biology and Department of Ecology, Evolution, and Behavior at the University of Minnesota. **Douglas W. Smith** is currently project leader for the Yellowstone Gray Wolf Restoration Project in Yellowstone National Park. **Daniel R. MacNulty** is assistant professor of wildlife ecology at Utah State University. **Robert K. Landis** is a freelance videographer for Yellowstone National Park and has produced several documentaries for National Geographic and others.

The University of Chicago Press, Chicago 60637
The University of Chicago Press, Ltd., London
© 2015 by University of Chicago
All rights reserved. Published 2015.
Printed in the United States of America

24 23   3 4 5

ISBN-13: 978-0-226-25514-9 (cloth)
ISBN-13: 978-0-226-25528-6 (e-book)
DOI: 10.7208/chicago/9780226255286.001.0001

A Note on Accompanying Videos by Robert K. Landis
Videos of wolf-prey interactions, by Robert K. Landis, are available to readers of the print book at the following URL and with these password credentials:
URL: www.press.uchicago.edu/sites/wolves
User name: wolves
Password: hunt2015
Readers of the ebook will find the videos embedded in the text.

Library of Congress Cataloging-in-Publication Data

Mech, L. David, author.
  Wolves on the hunt : the behavior of wolves hunting wild prey / L. David Mech, Douglas W. Smith, and Daniel R. MacNulty with supplementary online video by Robert K. Landis.
      pages ; cm
  Includes bibliographical references and index.
  "Videos of wolf-prey interactions, by Robert K. Landis, are available to readers of the print book at the following URL www.press.uchicago.edu/sites/wolves . . . with password credentials" — CIP data.
  ISBN 978-0-226-25514-9 (cloth : alk. paper) — ISBN 978-0-226-25528-6 (e-book) 1. Wolves—Behavior. 2. Wolves—Food. 3. Predation (Biology). I. Smith, Douglas W., 1960– author. II. MacNulty, Daniel R. (Daniel Robert), author. III. Title.
  QL737.C22M39982 2015
  599.77—dc23

2014038338

We dedicate this book to Thomas J. "Tom" Meier and Kevin Honness, outstanding colleagues, friends, outdoorsmen, and wildlife biologists. Tom worked with wolves in Minnesota, Wisconsin, Idaho, Montana, and Alaska, and many times recorded wolves hunting. He topped off his career as a National Park Service biologist in Denali National Park and Preserve from 1986 to 2012. Kevin made many observations of wolf-prey interactions as a volunteer for the Yellowstone Wolf Project from 1997 to 1998 and led the reintroduction of swift foxes to portions of South Dakota for the Turner Endangered Species Fund from 1998 to 2008. We thank both of these fine workers posthumously for contributing observations to this book.

Contents

PLATES FOLLOW PAGE 82

A NOTE ON ACCOMPANYING VIDEOS
BY ROBERT K. LANDIS
Videos of wolf-prey interactions, by Robert K. Landis, are available
to readers of the print book at the following URL and with these
password credentials:
   URL: www.press.uchicago.edu/sites/wolves
   User name: wolves
   Password: hunt2015
Readers of the ebook will find the videos embedded in the text.

Foreword

THE FOCUS OF this book, the prey-killing behavior of a large carnivore, provides rich material for us to ponder. Reported here are real-world accounts of gray wolves interacting with their prey—tracking them down, testing them, and trying to kill them. There is certainly drama here, and one can sense the violence as well as a timeless natural beauty in the behavior of predator and prey. Occasionally the wolves succeeded, and a lucky observer would gain a life-long memory. Whether an encounter is a success or a failure, of course, depends on one's point of view.

There is one case of failed predation I will probably remember for the rest of my life. It was spectacular not because it involved wolves or other large carnivores, and certainly not because of the location, a busy intersection in the town where I live. It was memorable because I'd never before seen these species in a predator-prey struggle, and the outcome was so unexpected. A short-tailed weasel had almost caught up with a red squirrel on the thick trunk of an isolated old elm tree. The squirrel was racing for its life round and round the tree trunk, with the weasel just inches behind it. I stood transfixed by this roadside scene, while cars by the dozen streamed by, drivers oblivious to the drama. After what seemed like minutes, but was probably just a fraction of one, the squirrel scampered up the tree trunk and out onto a branch. Surely, I thought, the weasel will simply trap the squirrel out on the end of that limb. However, by then the weasel was exhausted, and when the squirrel went up the tree the weasel curtailed the attack and went down. I'd seen many cases where moose seemed to wear out wolves in chases that went for miles, but I didn't expect that to happen when predator and prey were so similar in size and ability. From that one observation I learned that squirrels, as well as moose, have a life-saving anti-predator strategy.

So it has been for the authors of this book, who gained insight into the hunting behavior of wolves one observation at a time, decade after decade, with progress coming at unanticipated moments, when the stars (wolves and their prey) were all aligned. The authors are well qualified, each having had extraordinary opportunities to make detailed observations of wild wolves hunting and killing most of the prey species that support wolves in North America. Dave Mech has made a concerted effort, for over 55 years, to observe wild wolves hunting as many different prey species as possible. No one else has such a diverse basis for understanding what wolves do in the wild. The reintroduction of wolves into Yellowstone opened up a major window on wolf-elk interaction, as wolves were dropped into the largest elk herd in North America. Doug Smith was there when wolves were brought back, and he is still there over 20 years later, carefully documenting the ebb and flow of wolf packs and their effect on elk population dynamics. Dan MacNulty was able to figure out the challenging logistics of camping out in snow caves in Yellowstone's remote Pelican Valley among the bison, thereby discovering fundamental aspects of what it means for wolves to subsist almost entirely on this most difficult of prey. The experiences of these biologists may never be repeated by anyone. Mech's record would be difficult to equal, and the opportunities presented to coauthors Smith and MacNulty, coming on the heels of wolf reintroduction into Yellowstone, were unprecedented and may not return.

The first few observations recorded for wolves and a specific prey species are usually very informative, suggesting many features that may be similar to wolves and other prey. In my own case, having watched hundreds of instances of wolves hunting moose and maybe one to two dozen actual kills, one more observation of wolves chasing moose may have few memorable elements. I can understand the motivation of author Mech to observe wolves hunting prey that he hadn't seen previously. Twice I have had the good fortune to accompany him to Ellesmere Island, where I observed Arctic hares for the first time—these are large white hares that gather in groups of a hundred or more in a landscape that seems as barren as Mars. When they are frightened they typically run *uphill* and, as explained in this volume, their top speed matches that of wolves. I puzzled over this behavior for a while, until I saw a wolf try to do the same thing—the larger mass of a wolf forces them to a proportionately slower speed when going uphill, while the sprightly hare scampers off. This was a useful lesson in surface area–to-volume relationships.

The reintroduction of wolves into Yellowstone National Park in 1995 and 1996 was one of the most important developments in wildlife conservation of the twentieth century. Not only were wolves restored to a magnificent landscape where they had been killed off by the federal government in the 1920s, but a few dozen wolves soon also began to "manage" prey populations with scrutiny that we simply can't fathom. I joined the roadside throngs to observe wolves encountering elk and bison, something I'd never seen before—more memories. I learned how strong is the dislike of wolves by bison during an observation from the roadside in the Lamar River Valley. A wolf pack ran down and quickly dispatched a cow elk from a small herd. While the rest of the elk scattered and the wolves began to regroup and feed on their kill, a nearby herd of bison came thundering over and moved all the wolves away from the downed elk. For almost an hour the bison kept the wolves away from their kill. This single observation is rich with possibilities for interpretation. This book is packed with similar raw material.

There is much that biologists don't know about wolves, and maybe our ignorance even exceeds our knowledge (how would we know?). This is true even for those dedicated enough to spend decades studying wolves. One intriguing area about which we know little is the learning process that makes an efficient predator out of a naive young pup, albeit one with strong predatory instincts. We know that wolves can figure out how to kill prey given nothing more than instinct and opportunity, but what role do parents play in teaching hunting and killing behavior to their young? After all, it seems to take years for a young wolf to develop real prowess with large prey such as moose and bison. Mech made some fascinating observations of adult wolves at Ellesmere that seemed, indeed, to be actively teaching their young how to kill juvenile Arctic hares. Read on and be intrigued.

What does this book accomplish that can't be found in other books about wolves? It has rare breadth, presenting wolf-prey encounters across a diversity of prey that no single person has ever seen, even biologists with a lifelong opportunity to observe wolves in the wild. Many lifetimes, then, went into the making of this book. It is safe to say that there has probably never been a wolf that has seen everything represented here, as most wolves live out their lives within a few hundred miles of where they were born, with access to only one or two primary species of prey. If wolves could read, this might very well be their favorite book.

Rolf O. Peterson

Preface

DURING THE LAST several decades much has changed in the wolf's world, mostly for the better. The species—while inhabiting only Minnesota and Isle Royale National Park, Michigan—has progressed from being officially endangered in the 48 contiguous United States to being recovered and removed from the federal Endangered Species List in several northern states. Minnesota's wolf population increased from about 750 to 3,000 and proliferated into Wisconsin and Michigan, which now harbor over 1,200. Montana, Idaho, Wyoming, Washington, and Oregon now host more than 1,700 wolves. Other wolves have even ventured into California, Colorado, North Dakota, South Dakota, Indiana, Illinois, Missouri, and New York. Mexican wolf numbers now exceed 83 in Arizona and New Mexico. In Europe, the 100 wolves that remained in Italy in the 1970s have increased considerably and flowed over into France and Switzerland. Poland's have helped repopulate Germany, and Scandinavian wolves now number over 250.

Much research and public interest has accompanied these increases, and wolves have again gained both strong advocates and adversaries (Mech 2013). Advocates often cite what they feel are positive aspects of wolves' hunting behavior, such as evidence that these carnivores kill primarily individual prey that are old, sick, weak, or otherwise debilitated. Adversaries tend to emphasize what they consider negative traits: killing for sport, "slaughtering prey," and killing more than they eat ("surplus killing").

The above claims involve the wolf's hunting habits, and those habits have been the subject of scattered research and observation over the past several decades dating back to Adolph Murie's 1944 study *The Wolves of Mount McKinley*. Numerous accounts of first-hand observations of wolves hunting a wide variety of prey

have accumulated one-by-one. Many of these observations have been made from aircraft because, in much of wolf range, use of aircraft is the only way to observe a hunt continuously over its long distance (see, e.g., the video "Wolves Hunting Caribou" from the BBC program *Planet Earth*: http://www.youtube.com/watch?v=AoE6geAq1k8). In wide-open vistas such as the Arctic, observers have been able to describe entire hunts from the ground. Most such observations have been made opportunistically. However, more recently, the intensive ground and aerial surveillance of radio-collared wolves in Yellowstone National Park has allowed a new bonanza of wolf hunting-behavior observations, especially of elk and bison.

So far, no publication has assembled these diverse observations and tried to assess the commonalities. That is what we try to do in the present volume. We have gathered published first-person accounts of wolves hunting as many prey species as we could find and added our own unpublished observations made over a total of about 100 person years among the three of us. In any discussion of wolf hunting behavior, the possible use of strategy arises, and we discuss that in various chapters. By strategy, we mean the invoking of deliberate maneuvers and/or timing that tend to improve the chances of capturing a prey animal.

Almost all of this book's observations are of wolves in North America because that is primarily where most such wolf research has been conducted. However, there is no reason to believe that wolves hunting their sundry prey in Eurasia or the Middle East hunt any differently.

The observations of wolves hunting described in each of the following chapters were made by the authors of the publications cited or in a few cases by others whom the publication's authors cite, and the unpublished accounts were by those credited at the beginning of each account. Unless a description is in quotes, we have paraphrased it for brevity and to comply with a consistent style, including changing "alpha" to "breeder" or other appropriate synonyms in keeping with Mech (1999) and Packard (2003), while striving to retain all the relevant information. All distances included were estimates only, and where individual radio-collared wolves are numbered, their age and sex are given if they were the sole individuals involved in the hunt.

Because many more accounts exist of wolves hunting certain species of prey than others, we made no attempt to include all available accounts. Rather, in some chapters, especially those featuring prey for which large numbers of wolf hunting observations exist, we included the more informative and interesting observations by a variety of observers. In others, we included as many accounts as space would allow. Our intent is to immerse the reader in the wolf's life as this intrepid creature strives to find, catch, and kill its wide variety of prey.

Acknowledgments

L. David Mech prepared this book while employed by the US Geological Survey, D. W. Smith by the US National Park Service, and D. R. MacNulty by Utah State University. We thank the following for providing their observations of wolves hunting: C. Anton, C. Benell, A. Blakesley, M. Britten, K. Cassidy, B. Dale, T. Dawson, F. Dean, D. Dekker, P. Del Vecchio, N. Gibson, D. Hiser, N. Hoffman, K. Honness, S. Knick, M. Korb, A. La Fever, N. Legere, B. Looney, L. Lyman, H. Martin, R. McIntyre, M. Metz, R. Peterson, E. Proetz, B. Schults, E. Smith, D. Stahler, J. Tasch, T. Wade, and L. Williamson. In addition, we thank two anonymous reviewers for their helpful suggestions for improving the manuscript, as well as administrative assistant Nancy L. Anderson of the US Geological Survey for helpful secretarial support throughout the preparation of the manuscript. Kira Cassidy, Yellowstone Wolf Project, helped select and organize the elk photos.

Introduction

The Wolf as a Killing Machine

MANY PEOPLE—ESPECIALLY THOSE opposed to wolves—think of the wolf as a killing machine. And in a general sense they are right. The wolf has evolved to kill prey much larger than itself. Watching a wolf or a pack of wolves kill a deer or moose or elk, one is impressed with their prowess. Whether it be a wolf clinging to the nose of a moose or seizing the throat of an elk, one cannot help but watch in awe at the spectacle (www.press .uchicago.edu/sites/wolves, video 1). Even just finding the remains of a wolf-killed caribou or a domestic cow, with leg bones strewn, ribs chewed, and the stomach contents frozen in a pile on a blood-soaked mat of snow fills one with wonder—and some people with disgust. At first glance, it certainly appears that a wolf, or especially a pack of wolves, can kill just about anything it wants. In fact, many laypeople even believe that wolves kill for the sheer sport of it. The reality is different and far more interesting.

Several authors have discussed wolf hunting of specific prey species, as will be covered in each of the following chapters. Over a half-century ago, Young and Goldman (1944, 75) discussed the general hunting behavior of wolves with the succinct paragraph, "The hunting technique of wolves is based on the exhaustion of their prey. Their greater endurance over that of most, if not all, of the big game animals, is one of their main assets in overcoming prey." Since then, however, science has added many amazing specifics.

Murie (1944), Mech (1966a), Peters and Mech (1975), Allen (1979), Mech and Peterson (2003), and many others have described and discussed general wolf hunting behavior in detail, and Kunkel and Pletscher (2001), Kunkel et al. (2004), Hebblewhite et al. (2005), Sand et al. (2006), and MacNulty et al. (2007, 2009a, 2009b, 2012, 2014) have added considerably to the specifics of wolf hunting behavior. Whereas the earlier studies were primarily descriptive, the more recent investigations have been more quantitative, thus helping to dispel several of the earlier notions about the wolf as a killing machine. Here we synthesize this information and add to it as background to understanding the behavior of wolves as they hunt specific types of prey described in the following chapters.

Mech (1970, 193) began his general discussion of wolf hunting behavior as follows: "Much has been written about the hunting behavior of wolves, and they have often been credited with the use of teamwork and complex strategy in their quest for food. However, most conclusions about the hunting habits of the wolf have been based on hearsay, old published descriptions by nonobjective observers, and interpretations of tracks in the snow. Actual observations of wolves hunting prey are scarce, and accurate published accounts of these incidents are even scarcer."

Since that time biologists have many times observed wolves hunting deer, caribou, elk, bison, musk oxen, and Arctic hares. In this book, we combine the descriptions of those accounts with those published previously of wolves hunting Dall sheep and moose. Further, we highlight new information about wolf hunting behavior drawn from these accounts. Mech and Peterson (2003, 131) had the benefit of some, although by no means all, of the new data and came to view the wolf as "a highly discerning hunter, a predator that can quickly judge the cost-benefit ratio of attacking its prey. A successful attack and the wolf can feed for days. One miscalculation, however, and the animal could be badly injured or killed." Wolves have been killed by several prey species, including white-tailed deer, one of the wolf's smallest hoofed prey (table I.1). Thus the central problem for wolves on the hunt is to kill without being killed.

TABLE I.1.   Prey species known to have killed wolves in self-defense

Prey Species	Reference
White-tailed deer	Frijlink 1977; Nelson and Mech 1985; Mech and Nelson 1990
Moose	MacFarlane 1905; Stanwell-Fletcher and Stanwell-Fletcher 1942; Mech and Nelson 1990; Weaver et al. 1992
Musk oxen	Pasitchniak-Arts et al. 1988
Elk	MacNulty (2002); chap. 4
Bison	Chapter 6

Note: Based on 2,134 wolf skulls, 36% showed prey-caused injuries (Haugen 1987).

The problem of trying to kill without being killed or injured is common to most large carnivores because of the well-developed defensive equipment and behavior of most large prey (Mukherjee and Heithaus 2013). However, the extent of that problem varies with the species of those carnivores and of their prey. Generally the felids, except for the cheetah, tend to stalk or ambush their prey, thus singling it out and surprising it, whereas the canids tend to charge and chase their prey, singling out an individual during the chase. The stalkers and ambushers still must wrestle their large prey, but depending on both predator and prey species, the stalkers and ambushers tend to at least stop prey in place and, using sharp claws as well as fangs, start the killing process well fastened to the prey. The large cats apply a killing bite and can then eat their prey after it is dead.

In contrast, wolves, as well as African wild dogs and hyenas, must catch up to their fleeing prey, slow it down, chew through thick hair and hide, and eventually disable the struggling prey well enough to start feeding. Or with prey such as moose, bison, and larger elk that often stand their ground, wolves must confront them and attempt to bite into them until they are incapacitated. Because of this prolonged struggle, wolves are much more vulnerable to injury or death by their prey than the felids.

Whether wolves' ability to size up the cost-benefit ratio of each hunt is innate or learned is unknown, but the animals seem to be able to assess which prey species in a given multiprey system at a specific time are the most profitable to hunt with a minimal risk of injury (Weaver 1994). The basic hunting, attacking, and killing tendencies of wolves, like those of most predators, are innate (Fox 1969), of course, just as with domestic dogs and cats. The extent to which learning helps refine wolf hunting behavior is an interesting question that we are finally beginning to answer (Mech 2014). Coauthor D. W.

Smith has observed wolf pups involved in hunts all winter long "rarely doing anything," although quite possibly they are still learning something about hunting behavior. In a thorough investigation of factors influencing wolf hunting ability of elk, MacNulty (2007, 69) concluded that "the development of hunting ability likely consists of an early phase (< ~ 2 years) where learning is most influential, and a later phase (~ 2–4 years) where increasing [wolf] mass is most influential." It might not take too many close calls while a wolf attacks a prey animal to teach the wolf that some prey are better left alone. At the same time, too apprehensive a wolf could quickly go hungry. One way or another most wolves find that fine balance between being too aggressive and too timid— or they die.

Before wolves are faced with that dilemma, however, they first must solve two other problems: finding a prey animal and then confronting it. Some prey run as wolves approach, some stand their ground, and some run and then stand their ground (www.press.uchicago .edu/sites/wolves, video 2). Most often deer and caribou tend to flee when wolves approach, although even deer sometimes stand their ground. Larger prey usually stand their ground. For example, about 55% and 79% of the time elk and bison, respectively, tend to confront wolves (MacNulty 2002). With 180 moose on Isle Royale, 63% ran immediately, 24% stood as wolves approached them, and 13% fled but stopped when wolves caught up to them (chap. 2). Of the 63% that ran, seven were killed (Mech 1970; Peterson 1977). Only when confronting their quarry must wolves decide whether to risk trying to kill it. With small prey such as hares, beavers, or geese, the risk is nil. But with most others the risk is great. This claim of ours may surprise even some fellow scientists because of the recent notion that prey live in fear of wolves (Brown et al. 1999; Berger 2008). On the contrary, most prey present great challenges to wolves (www.press.uchicago.edu /sites/wolves, video 3).

The wolf's physical abilities help the animal meet these challenges. Mech (1970) and Peterson and Ciucci (2003) discussed these characteristics in detail. To start with, wolves are the largest members of the canid family. Adult males average 28–45 kg and sometimes reach 78 kg, while females average 25–39 kg and can reach 54 kg, depending on the area. The feet of wolves are also large, and they are blocky but flexible, which allows wolves to tread on many types of terrain, including rough, rocky areas, and the wolf's legs are long, propelling them at high speed and facilitating extensive travel. Wolves' usual rate of travel is 8.7 km/hr (Mech 1994b), and they can cover 76 km in 12 hr (Mech and Cluff 2011).

The wolf's ability to travel is critical to solving the problem of finding prey and has been much touted by previous authors. Mech et al. (1998) even dubbed a chapter of their *Wolves of Denali* book "The Wolf Is Kept Fed by Its Feet," after an old Russian proverb. The ability to travel far, wide, and long not only helps wolves find prey animals but, just as important, it helps them find specific individuals they can catch with a minimum of risk (see chap. 9). With experience, they learn the best areas to travel to, including habitat patches and physical features favorable to success, thus increasing their efficiency. For example some 5–10 yr after wolves recolonized northwestern Montana, they tended to travel (during winter) through areas with deer, elk, and moose densities 5.3, 10, and 40 times higher, respectively, than experimental control routes (Kunkel and Pletscher 2001). In Yellowstone, flat grasslands near roads and streams were more favorable to wolf hunting elk (Kauffman et al. 2007).

It might take wolves hours of traveling to locate prey, depending on the area and prey species, and usually it takes many hours or days for wolves to find prey they can kill (Mech 1966a; Peterson 1977; Kunkel et al. 2004). A notable exception was in Yellowstone National Park during the first few years of wolf restoration. An especially high number of elk lived there, including many vulnerable ones, and it only took about an hour for wolves to find and kill one (MacNulty 2002).

Some prey may live more or less singly, like moose or like deer in summer, while several wolf prey species group up in herds of up to thousands, like caribou. Herding has many antiwolf advantages, which we will discuss in later chapters, although under some circumstances smaller groups (specifically with elk) might be more advantageous to the prey species (Hebblewhite and Pletscher 2002; Creel and Winnie 2005). Wolf travel habits must vary to meet each type of challenge. Each time the wolves' travels take them near a prey animal, the predator's sharp senses come into play. Wolves use all their senses to locate prey, depending on the circumstances. On Isle Royale in Lake Superior, for example, wolves used scent to locate moose 10 of 17 times, vision six times, and tracking in snow once (Peterson 1977). They can see prey at least as far as humans can (Mech 1988, 2007b) and can smell prey as far as 2.4 km (Mech 1966a) and possibly much farther. One wolf traveled more or less straight to a herd of caribou some 103 km away (Frame et al. 2004). Although there's no way of knowing whether the wolf smelled the herd at that distance, caribou drift around a great deal, so memory alone probably did not account for the wolf's locating the caribou. Even humans can smell odors of such things as forest fires, paper mills, and so forth for hundreds of kilometers when downwind. Thus it does not seem unreasonable that a wolf, whose olfactory sense is probably at least as good as that of a dog—that is, 100–10,000 times more sensitive than that of a human (Asa and Mech 1995)—could detect the direction of a caribou herd many kilometers away by catching a few molecules of its odor.

Once wolves encounter quarry that flee, then the chase begins. Generally the sprint speed of prey is tightly related to the vulnerability of the prey to their main predators (Bro-Jørgensen 2013). Wolves can run for several kilometers at 56–64 km/hr (Stenlund 1955; Mech 1970), and it is not unusual for wolves to chase prey for more than a kilometer (table I.2). The longest pursuit by wolves ever recorded seems to be that of a wolf chasing a caribou for 8 km without catching it (Crisler 1956), although a single wolf once chased, tracked, and followed a deer 20.8 km, with an unknown outcome (Mech and Korb 1978). One 11–13-yr-old wolf chased an arctic hare for 7 min before catching it (Mech 1997).

Most wolf prey try to run to any nearby water, where, if the water is deep enough, they can often escape or better ward off the wolves. However, as will be apparent in chapters 1, 2, and 4, they will not always be successful (www.press.uchicago.edu/sites/wolves, video 4).

TABLE I.2.   Distances wolves chased various prey

Prey	Chase Distances	References
Caribou	8 km, longest	Crisler 1956
Elk	351 m, average; 1,031 m, longest	Carbyn 1974
Elk	115 m, average, successful	Paquet 1989
Elk	260 m, average; 10–1,700 m	Wilkenros et al. 2009
Elk	978 ± 142 (SE) m	Kaufmann et al. 2007
Elk	2.4 km, longest	Cowan 1947
Moose	4.8 km, longest	Mech 1966a
Moose	883 m, average, successful	Paquet 1989
Moose	0–1.7 km, successful	Wilkenros et al. 2009
Moose	5 km, longest, failed	Wilkenros et al. 2009
Roe deer	0–2.3 km, successful	Wilkenros et al. 2009
Roe deer	13.7 km, failed	Wilkenros et al. 2009
White-tailed deer	average 600 m–1.9 km, failed; 2.1 km, successful	Kolenosky 1972
White-tailed deer	5.6 km, longest	Kolenosky 1972
White-tailed deer	159 m, average, successful	Paquet 1989
White-tailed deer	20.8 km, longest	Mech and Korb 1978
Deer, elk, sheep	5–1,000 m	Huggard 1993a

On catching up to or confronting one of their usual prey species (a hoofed animal rather than smaller prey), wolves' 6 cm long fangs come into play, powered by extremely strong masseter muscles (Peterson and Ciucci 2003), with a bite force of 28 kg/cm^2 ("Dangerous Encounters: Wolf Bite Force," *Dangerous Encounters with Brady Barr*, May 14, 2007, http://channel.nationalgeographic.com/wild/dangerous-encounters/videos/wolf-bite-force/). Wolves can break open the skulls of many of their prey animals, including that of an adult cow moose (Mech unpublished), and can shear off a domestic cow's tail (Young and Goldman 1944). MacNulty has watched wolves pull the tails off bison. The wolf's general strength also contributes greatly as the predator leaps at its quarry and tries to tear at it, as will be apparent in the following chapters. R. O. Peterson watched a wolf cling to the hind leg of a moose while being dragged for dozens of meters (Peterson and Ciucci 2003). Coauthor D. W. Smith watched a wolf grab a bull elk's nose and get dragged by the elk while it ran over a downed log, after which the wolf let go.

Added to these basic individual qualities is the common pattern of more than one wolf making the attack. Single wolves have been recorded killing each of the common prey species including musk oxen, bison, and moose (Mech 1970). However, more often wolves hunt in packs, especially during winter when even the young of most of their prey have grown considerably. Hunting as a pair or a pack, of course, offers a considerable advantage to the wolves.

The claim has been made that, in response to increases in winter snow, wolves increase their pack size to hunt in larger packs and thus kill more prey (Post et al. 1999). However, that is not the case. Wolf-pack hunting size is a function of the survival of pups born the previous summer, adult survival, and rate of dispersal (Mech et al. 1998). Increased snow brings greater prey vulnerability, which fosters pup survival and decreases dispersal, thus allowing pack size to build up (Peterson and Page 1988). Larger packs in winters of more snow, therefore, are a consequence of higher kill rate, not a cause of it (L. D. Mech and R. O. Peterson, unpublished). Although two or more wolves attacking a prey animal are clearly more effective than one, it is not necessarily true that the larger the pack, the more effective it is. That relationship may seem logical, but biologists long ago began to question it. Mech (1970, 42) noticed in his Isle Royale studies of 15 wolves hunting moose that "seldom was the whole pack in on the kill; usually only five or six animals actually made contact with the prey." Since then, other workers have found that the larger the pack, the less food obtained per wolf (Thurber and Peterson 1993; Schmidt and Mech 1997). More recently, MacNulty et al. (2012) showed that, with wolves hunting elk, hunting success peaked at four wolves, a result much like that found in simulated hunts using groups of robots (Lerman and Galstyan 2002), and that, with wolves hunting bison, hunting success increases across larger pack sizes (MacNulty et al. 2014). These findings fit well with the composition of wolf packs, almost all of which are composed of a pair of breeding adults (formerly called alpha pair [Mech 1999, 2000]) and their offspring from the previous 1–3 yr. Thus most packs, regardless of size, would consist of at least two large, experienced, mature wolves plus any number of "apprentice" offspring. Maturing 2- or 3-yr-olds (usually in larger packs), which would be the most vigorous pack members, join the adults and assume the more strenuous tasks. The most recent offspring would be the previous year's inexperienced pups, which, it is not surprising, are usually "free riders" in the wolf-prey interaction, although many adults also free ride at times (MacNulty et al. 2012).

Thus, hunting success peaks in wolves aged 3–5 yr (Sand et al. 2006; MacNulty et al. 2009a), and male wolves of this age tend to be the most effective hunters because they are largest (MacNulty et al. 2009b). Males in general are better hunters than females because males are larger, usually 9%–32% heavier than females (Butler et al. 2006; Mech 2006; MacNulty et al. 2009b), and outperform females by 42% in attacking, 43% in selecting, and 78% in killing prey (MacNulty et al. 2009b). In small packs, wolves 3–5 yr old are the breeders, whereas larger packs also include older offspring (Mech and Boitani 2003). Mech has observed younger wolves performing some of the most arduous tasks during an attack, while the breeders then come in to help make the kill. With arctic hares, yearling wolves sometimes did the chasing, and the resting parent would then grab the hare as it passed. With elk and musk oxen, the younger wolves slowed the prey, and the parents joined in for the kill. These findings about ages of the best hunters mesh well with those showing that in a natural (unhunted) wolf population, the average length of time adult wolves spend in their pack is only 4 yr (Mech et al. 1998). In fact, in Yellowstone, where we have the best data, the median life span of wolves is 5.94 yr (MacNulty et al. 2009a) and outside of the park, only half that (coauthor D. W. Smith, pers. comm.).

Besides the wolf's physical adaptations for hunting large mammals, wolves are thought to possess several behavioral/psychological/cognitive characteristics that also adapt the animals to their hunting lifestyle, among them persistence, coordination, cooperation, concealment, strategy, and cognitive mapping (Clark 1971; Peters and Mech 1975; Mech 2007b). Wolves are excellent learners and enjoy strong cognitive functioning. Several experimental studies show that wolves learned various tasks at least as well as domestic dogs and, with several tasks, better than dogs (Packard 2003). Since then, sophisticated tests of wolf cognition, for example, evince that wolves are excellent at using "gaze cues" to focus their attention where others are looking (Range and Viranyi 2011). This trait would certainly serve them well in hunting along with their associates, and Clark (1971, 146) observed that "coordination among wolves under hunting circumstances is apparently by visual cues" and that wolves do not vocalize while hunting. In addition, wolves show greater insight than do malamutes during standard lab testing (Frank et al. 1989) and sometimes seem to exhibit foresight, understanding, and ability to plan as part of their hunting strategy (Mech 2007b). The use of any of these traits during hunting would be strongly selected for by the animals' need to kill without being killed.

The questions of how much strategy (or tactics) wolves use in hunting, and which strategies, have been discussed for decades. Olson (1938), Rutter and Pimlott (1968), Kelsall (1968), Clark (1971), and Haber (1977) thought that wolves use ambushing, while Olson (1938) and Kelsall (1968) suggested they may also use relay running. Whereas there is little evidence of relay running, coauthor D. W. Smith, who has viewed many hunts, believes that from hunt to hunt individual wolves in large packs might differ in the extent of their participation, thus spreading risk. Although there is no quantitative evidence for this idea (MacNulty et al. 2012), Coauthor D. W. Smith has noted that in some hunts, certain big males are missing, whereas in the next hunt they are involved.

It has been said that wolves hunting deer use a line-abreast formation (Stenlund 1955), but Mech who has watched several wolf hunts of deer has never seen that tactic nor has it been reported by anyone else. Still, wolves sometimes do approach caribou (Banfield 1954) and musk oxen (Gray 1983) in this manner. Murie (1944, 110) hinted at wolf use of a simple tactic when he described how wolves hunt Dall sheep (chap. 5): "The method is to get above a sheep and force it to run down." Bibikov (1982, 128) speculated that "wolves may appraise the endurance of a potential prey and the capacity for defense by whether it flees upwards or downwards. A healthy and strong animal is capable of fleeing upward, whereas a sick or physically impaired individual is less capable of fleeing up a slope and therefore may be more likely to run downward." Wolves often killed deer on lakes during winter in Minnesota (Mech et al. 1971) and red deer and wild boar in creek beds and ravines (Gula 2004).

Kelsall (1968, 252–53) concluded that wolves used three main strategies to catch caribou: (1) "When caribou are dispersed singly, or in small bands, wolves generally use stealth and ambush in hunting. . . . A single wolf killed a calf, from a small band of caribou, by lying in wait on a riverbank. The wolf remained hidden until a small band of caribou walked directly below. It then dashed directly downhill at the animals and managed to single out a 2-month-old calf and edge it into the river. The wolf ran no more than 60 meters in total, and caught and killed [it] before the calf could get into deep water. Of several instances where wolves have been seen to lie in wait, they have invariably taken advantage of situations where they would have a hidden and downhill start." (2) "Relay running of caribou by 2 or more wolves is commonly reported as being a successful killing technique. This has not been observed by biologists, but there seems no reason to doubt that it takes place. One or more wolves will chase caribou, while 1 or more loiter to the rear, taking advantage of every turn of the chase to save time and energy." (3) "Most caribou are probably taken by wolves from bands sufficiently large that they hinder each other in running."

Haber (1977) believed that wolves used the following strategies for catching caribou: (1) deploying at least two individuals around the targets or along an escape route presumably in ambush; (2) maneuvering prey into terrain advantageous to wolves, and (3) using a decoy wolf to distract a caribou while a second wolf sneaks closer from another direction. Haber's hunting accounts 31–33 in the caribou chapter illustrate these alleged strategies.

How valid are the above generalizations about wolf use of more complex strategies to capture caribou? Mech (1970) seemed to have accepted Kelsall's (1968) claims of wolf strategies uncritically except for that of relay running. If wolves do employ such higher-order techniques to hunt caribou, presumably they also at times use them on other prey. However, Peterson and Ciucci (2003) noted that several biologists have reported many other

observations of wolves hunting various species without mentioning such cooperative maneuvers. To assess the opinion of contemporary wolf biologists, Mech sent a questionnaire to 19 wolf biologists to determine their views about whether wolves use such complex strategies to obtain prey. A summary of the results of that survey follows: "A survey of 19 biologists who have observed many wolf-prey encounters reveals no unanimity in their beliefs about the wolf's possible use of ambushing or relay running. Most wolf biologists believed that wolves do sometimes use some forms of cooperative strategy, but the number of descriptions [they gave] including convincing examples of it is low. Most described chases were simple and straightforward" (Peterson and Ciucci 2003, 122). However, we must add that 15 of the 19 respondents believed that wolves use some form of cooperative hunting strategy, and two respondents replied "maybe." Regardless, quantitative analysis has yet to demonstrate coordinated behavior.

The problem with trying to document and quantify strategic cooperation when watching wolves hunt various prey is that each such observation could be considered a type of Rorschach test. Even when a hunt is observed in full, there are often so many animals in play, both wolves and prey, that it is possible to interpret what could be random, or at least independent, movements as a pattern. It would take observing a large number of hunts with similar patterns to establish that indeed a certain pattern did amount to a cooperative strategy. Furthermore, certain key aspects of wolf hunting behavior that appear as use of strategy follow simple rules that even robots or agents on a computer can use to produce "complex and seemingly intelligent and purposive behavior (Braitenberg 1986; Mataric 1992)" (Muro et al. 2011, 193).

When one peruses all the various hunting accounts we describe in this book, no strong patterns seem to show up consistently. With arctic hares, Mech observed hunts in which yearling wolves chased the hares while adults positioned themselves and ambushed each hare as it came by, and he was convinced this behavior was deliberate (chap. 8). Mech (2007b) also described wolves waiting hours in a place where he successfully predicted the wolves would ambush a musk ox herd, as well as an incident where four members of a pack waited in one place while two others stalked a musk ox herd from another spot, and all attacked when the herd spooked. Wolves generally do seem intelligent enough (Packard 2003) that it would not be surprising that packs whose members have associated and hunted together long enough would almost automatically learn how to maneuver relative to one another in ways that increase their chances of success (www.press.uchicago.edu/sites /wolves, video 5). The extent to which this behavior amounts to deliberate strategy will probably be debated for decades more.

Any type of advantage a wolf can gain either as a single individual or as a pack member serves it well because most of its usual prey are creatures much larger than themselves. Wolves eat just about any kind of food, animal or vegetable (Peterson and Ciucci 2003), but they prefer, and most often consume, large hoofed mammals such as deer, moose, elk, caribou, wild boar, bison, and the like, depending on where they live. Domestic livestock also fit this bill. The wolf's smallest consistent prey in certain regions are hares (Tener 1954a). Contrary to the popular book (and movie) of fiction *Never Cry Wolf* by Farley Mowat (1963), nowhere do wolves subsist on mice.

All the hoofed creatures except the domestic variety possess alert senses and dangerous defenses. Most are neither easily found nor easy to catch and kill. They each have their specific defenses and antipredator behavior, which we detail in the following chapters. Suffice it to say here that all the wolf's hoofed prey species can be lethal to wolves (fig. I.1; table I.1). It would be dangerous and foolhardy for wolves to "kill for sport."

That wolf prey are not easy to catch and kill is hard for many people, including some ungulate biologists, to appreciate. Because the wolf possesses fangs and usually lives in packs, people often believe that wolves can overcome any prey they encounter. However, generally that is not the case. Hunting success rates for wolves are usually low for all ungulate prey species (Mech and Peterson 2003). True, after a hard winter with deep snow, wolves might have an easier time killing prey come spring not just because the deep snow might hinder prey escape but also because prey are weakest at that time. (Under such conditions wolves sometimes kill more prey than they can eat at the time, "surplus killing." We will cover this subject later.) But those severe conditions come few and far between, and they only occur during part of the year.

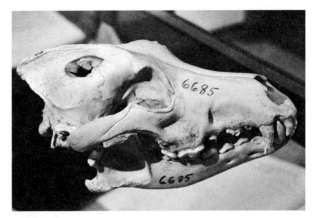

FIGURE I.1. Wolf skull with cranium pierced by white-tailed deer hoof (Mech and Nelson 1990).

Wolves must make it through the entire year and gain enough extra energy to raise their pups each year.

In addition to the specific defenses each prey species possesses, at least some, and probably all, of these prey can distinguish the odor of predator urine or feces (Muller-Schwarze 1972; Steinberg 1977; Ozoga and Verme 1986). There is still much to learn about prey response to predator urine and feces odor (Apfelbach et al. 2005; Berger et al. 2001), but it is reasonable to suggest, as Adams et al. (1995) did, that this ability may help prey to frequent areas with lower wolf densities, one more way of helping to thwart wolves.

By way of contrast, not all prey species of wolves necessarily maneuver to inhabit areas with fewer wolves, especially species that have effective defense mechanisms (Mao et al. 2005; Wirsing et al. 2010). Examples are deer and elk in Central Ontario and elk in Yellowstone (Mao et al. 2005; Kittle et al. 2008). It appears that in such a situation factors other than wolf density affect the type of areas prey frequent, a tribute to their antipredator adaptations.

In the "arms race" (Dawkins and Krebs 1979) between prey trying to survive while wolves try to kill them, prey animals respond to wolf presence by increasing their vigilance (Huggard 1993b; Laundre et al. 2001; Childress and Lung 2003) and frequenting habitat of increased structural complexity (Atwood et al. 2007). Prey that are nomadic such as caribou and musk oxen no doubt move out of areas where they have been attacked by wolves (Charnov et al. 1976). Elk certainly do (Gude et al. 2006).

It appears that the way wolves respond to maximize

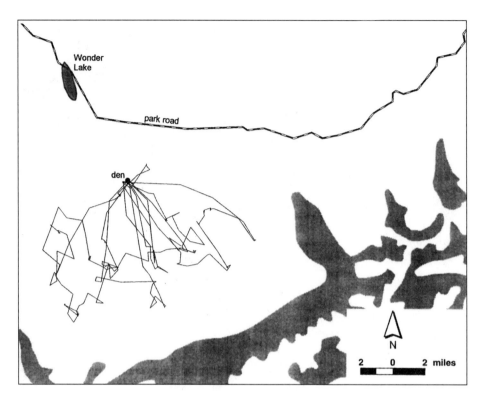

FIGURE I.2. Alternate hunting routes of a 4-yr-old male wolf from a den in Denali National Park over a period of 2 wk (Mech et al. 1998).

their chances of encountering the least wary prey is by constantly altering the areas in which they hunt, as Carbyn et al. (1993) surmised. When headquartered at a den, they head off each day in a different direction as evidenced by adult male Wolf 9628 estimated at 4 yr old in Denali Park (fig. I.2). During the rest of the year, wolves tend to rotate the use of their territory. In Poland, wolves tended to revisit each part of their territory about every 6 d on average (Jedrzejewski et al. 2001), and in Minnesota during summer, they use different parts of their territories on consecutive days. Mean daily range overlap by each individual wolf in a pack during summer was only 22% (Demma and Mech 2009a). Some trappers and early researchers might even have intuited such wolf movements. Olson (1938, 332) stated that wolves "have a beat which they cover every two or three weeks and a trapper who knows the route of a pack can bank on the possibility of its appearance in a certain locality regularly."

Wolves also seek to surprise their quarry. They sneak up when possible even in open terrain like tundra (see, e.g., "Musk Ox Save Calf from Wolves," an episode of *Frozen Planet*: http://dsc.discovery.com/videos/frozen -planet-musk-ox-save-calf-from-wolves.html). And they

seem not to be interested in prey that happens to stumble into them. That is, their strong tendency is to approach their quarry on their own terms. Long ago, Crisler (1956, 340) observed that "a wolf prefers not to be eyed when approaching its prey." Similarly, Kunkel and Pletscher (2001,526) concluded "the element of surprise (as provided by stalking cover) appears to be a very important factor affecting predation success of wolves in our study." Surprise as a valuable aspect of wolf hunting behavior seems perfectly logical, too, because that tactic would give wolves their maximum advantage. Nevertheless, wolves hunting musk oxen in some areas (chap. 7) and hunting elk (chap. 4) and bison (chap. 6) in Yellowstone commonly approached them with no attempt at concealment (Smith et al. 2000, MacNulty 2002).

The stages of a wolf hunt can be viewed in various ways. Generally, predatory behavior was traditionally divided into search, pursue, and capture (Holling 1965, MacArthur and Pianka 1966). However, MacNulty et al. (2007) reviewed the literature on large carnivore hunting behavior and found wide disparity in definitions of hunt stages. For example, Mech (1970) broke wolf hunts into the following phases: the stalk, the encounter, the rush, and the chase, whereas after recording ≥2,000 hr of wolves hunting elk and bison, MacNulty et al. (2007) proposed parsing hunting behavior into search, approach, watch, attack-group, attack-individual, and capture. These workers tested this more refined breakdown on seven other species of large carnivores—and five wolf populations other than the Yellowstone wolves on which they based their research—and proposed it as a common ethogram for discussing large carnivore hunting behavior. It worked well as a unifying construct that would then allow better comparisons of hunting behavior among various species and populations. This ethogram even worked well as a basis for designing a system of robots to emulate wolf-elk interactions (Madden et al. 2010).

The search part of the hunt occurs almost anytime wolves are traveling (Mech 1966a, 1970). They constantly sift their surroundings for the sight, smell, or sound of prey as they make their way around their territory, and any prey they detect might become fair game. At times while searching, wolves approach prey unknowingly, as when wolves detect that a prey animal—such as a deer in a thick, forested area—lies ahead but haven't pinpointed its location. Or when wolves approach an area where experience tells them prey reside such as a beaver pond or where young hares hang out (chap. 8). In these cases, the search phase may involve merely advancing toward the area. In any case, there may or may not come a time when the wolves merely watch their prey or watch for them. With prey species in open areas, such as caribou, bison, or musk oxen (chaps. 3, 6, 7), wolves may watch and study their quarry. With beavers and hares, and possibly mountain sheep, they may just watch until prey are in a vulnerable position. The predators then charge ahead, the phase that Mech (1970) described as the rush and MacNulty et al. (2007) as the attack phase. In herding species, this is the attack-group phase, which naturally then transitions into the attack-individual phase. Both types of attack can involve chasing or lunging, depending on whether prey flee or confront wolves, respectively (MacNulty et al. 2007).

The attack-individual phase of the hunt is the make-or-break stage. This is where the attacker must critically size up the situation. The wrong assessment, and the wolf or wolves end up either wasting considerable time and energy or getting clobbered by their quarry (ta-

ble I.1; www.press.uchicago.edu/sites/wolves, video 6). Interactions with prey at this time are quite tiring. Even when just chasing a hare, a wolf might rest for several minutes between catching one and eating it (chap. 8). With larger prey, wolves may wound the animal and then lie around resting or sleeping for hours before attacking it again to finish it off (chaps. 2 and 6). No doubt these considerations are why wolves have evolved a finely honed sense of caution and judgment as to whether to continue a hunt. Crisler (1956, 340) remarked on "how quickly the wolves had judged when a chase was useless." This phase of the hunt either ends in failure or the capture phase.

Most often the hunt ends in failure (see each chapter). Depending on the species, the season, the area, and the circumstances, wolves usually have a difficult time catching prey. Nine of 14 studies reported wolf hunting success rates of 1–9%, and the others from 13% to 56%, based on number of individual prey involved (Mech and Peterson 2003). Based on number of groups of prey attacked, success in killing a single individual varied from 10% to 49%.

Thus wolves literally lead a feast-or-famine existence. However, they are well adapted to this type of lifestyle. The longest reported period for which a wolf was known to have gone without eating was 17 d (Makridin 1962), but a dog once survived after having fasted for 117 d and losing 63% of its weight (Howe et al. 1912). When wolves do get a chance to eat, they can consume up to 10 kg at a time (Mech and Boitani 2003). A yearling female wolf left illegally in a trap for 5–7 d gained 19% of her original weight when fed and watered over the next 8 d (Mech et al. 1984). Wolves unsuccessful at hunting for too long can also rely on retrieving caches or scavenging to keep them going (Mech 1970; Peterson and Ciucci 2003).

This grand process of wolves hunting prey, whether the prey be hares, deer, caribou, bison, or any other species, is but one of a myriad of complex phenomena in the total natural scheme. However, it is the key to the wolf's survival, and for anyone interested in the wolf, it is a fascinating process. Understanding the reality of wolf hunting behavior, especially the numerous limits on the wolf's predatory ability, is vital to understanding the wolf and its interactions with prey. In the following chapters, we present the details of the behavior of wolves hunting their various prey as well as the defensive behavior of the prey.

1

White-Tailed Deer

TO THE PUBLIC, and even to some wildlife biologists, the white-tailed deer is seen as easy prey for wolves. This view probably results from the fact that deer represent one of the smallest of wolf prey and because wolves are large and live in packs. Compared superficially with moose or bison, deer would seem to be much easier prey. However, to capture and kill deer, wolves must first find them, catch up to them, and confront them. At each step in this process deer possess effective antipredator strategies. These traits help explain why deer and wolves can continually coexist without either one going extinct (Olson 1938; Stenlund 1955; Mech 2009), except under unusual circumstances (Mech and Karns 1977; Nelson and Mech 2006).

Science may know more about the interactions between wolves and white-tailed deer than it does about wolf relations with any of its other prey, except moose and elk. This is the case for two main reasons. First, some of the earliest studies of wolves were conducted where white-tailed deer were the wolf's main prey (Olson 1938; Stenlund 1955; Mech 1966b; Rutter and Pimlott 1968; Pimlott et al. 1969; Mech and Frenzel 1971). Second, much of the information results from the only long-term, intensive study that has employed radio tracking of both wolves and their prey. That study of wolves began in 1968 in the Superior National Forest of northeastern Minnesota, and incorporated radio tracking of deer starting in 1973. Through 2006, this study had livetrapped 712 wolves and radio-collared most of them, and through 2007, radio-collared 347 deer (Mech 2009). Thus numerous publications have synthesized and discussed wolf behavior hunting deer (Mech 1970, 1984; Mech and Frenzel 1971; Nelson and Mech 1981, 1993), and many others have synthesized information about wolf-deer interactions in general (Kolenosky 1972; Fritts and

Mech 1981; Kunkel and Pletscher 1999; Theberge and Theberge 2004; DelGiudice et al. 2009; Ballard 2011).

Here we highlight the major findings from the above studies of wolf hunting behavior and update them with information published since and with additional information gleaned from the many unpublished accounts and observations below. Unless otherwise mentioned, information in this chapter applies only to white-tailed deer.

In current wolf range, white-tailed deer are distributed throughout the Southwest, Midwest, and West of the United States and throughout southern and northwest Canada, with densities lower in extreme northwest Canada. Thus wolves prey on deer in many areas. That deer have been able to extend their range farther and farther north in the face of wolf predation (Heffelfinger 2011) is evidence of the species' considerably effective antipredator traits and behavior.

Adult female deer that live in wolf range generally weigh 70–80 kg (Mech and McRoberts 1990), and males weigh up to 180 kg (Sauer 1984). Only males possess antlers, which are fully hardened in fall and retained for a few months of fall and winter. All deer except newborns sport small, sharp, pointed hooves, which can kill a wolf (Frijlink 1977; Nelson and Mech 1990) or can alert other deer to danger (Caro et al. 1995), and all possess a conspicuous fanlike tail that is white underneath and that the deer "flags" or raises when disturbed.

Most deer occupy home ranges of about 0.4–7.0 km², with most females remaining close to where they are born while males disperse (Nelson and Mech 1981). Females generally produce their fawns relatively close to where they themselves were born, and a matriarchal society develops around them with offspring home ranges often overlapping those of their mothers (Nelson and

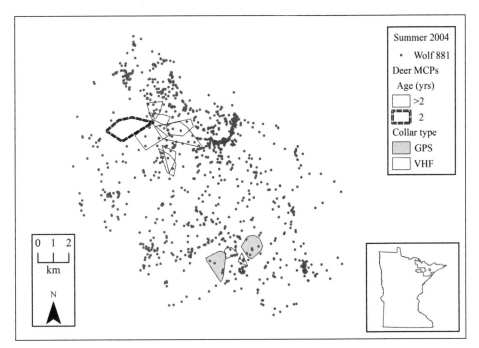

FIGURE 1.1. Home ranges of radio-collared deer in a radio-collared wolf-pack territory in the Superior National Forest of Minnesota. The dots represent the extent of the summer wolf-pack territory, and the outlines represent deer home ranges. Only a small proportion of the deer in the territory were radio-collared (Demma and Mech 2009b).

Mech 1981). Thus many deer home ranges fit within a single wolf-pack territory (fig. 1.1). Like most northern ungulates, deer begin gaining weight in spring and reach their peak weight in midautumn and their nadir in late winter or early spring (DelGiudice et al. 1992).

The weight and nutritional condition that deer lose throughout winter is directly related to snow depth and temperature (Verme 1963). Further, if winter conditions are extremely poor (deep snow primarily) or favorable, there is evidence that the extreme effect (positive or negative) on an individual deer's nutritional condition can accumulate over consecutive winters (Mech et al. 1987). Thus a series of severe winters could result in deer of poor condition, or vice versa. (This finding has since been both challenged [Messier 1991, 1995; Garroway and Broders 2005] and supported [Feldhamer et al. 1989; McRoberts et al. 1995; Post and Stenseth 1998; Patterson and Power 2002]).

There is also evidence that the extreme effects of snow depths can even persist across generations of deer, such that deer whose grandmothers lived through a winter of deep snow while gravid with offspring produced fawns whose own offspring (the F_2 generation) were in poorer condition regardless of the condition of the mothers (i.e., the F_1 generation [Mech et al. 1991]). Although this finding seems surprising, it was confirmed by Monteith et al. (2009) for deer and by Lumey and Stein (1997) and Bygren et al. (2001) for humans. (A possible mechanism is epigenetics [Pembrey 1996]). Of course, a prey animal's condition is extremely important to escaping predators.

In much of their northern range, many deer migrate between their summer ranges and winter "yards" or areas where deer from as far as 40 km away (Nelson and Mech 1981) concentrate. Usually by late winter most deer have migrated to winter yarding areas if snow is deep enough and/or temperatures low enough (Nelson and Mech 1981). In spring, deer migrate back to their summer ranges as snow melts and temperatures rise.

At least one important function of this yarding is for each individual to reduce its risk of predation. Nelson and Mech (1981, 36) explained this functioning as follows:

Yarding behavior provides several significant antipredator benefits: (1) The mere congregation of many animals creates a system of trails which can become escape routes during chases by predators. (2) Herding provides greater sensory capability and makes gregarious animals less vulnerable than solitary individuals to undetected predator approach (Galton 1871, Dimond and Lazarus 1974, Treisman 1975, and references therein). (3) Social grouping may confuse the search image of predators (McCullough 1969). One way such confusion may operate is suggested by observations of 15 deer groups in our study area chased by wolves in winter (Mech unpublished). The groups tended to split up when closely pursued, just as moose do (Mech 1966[a]). Such a maneuver could give group members an added chance of survival as the wolves tried to choose which individual to chase. (4) Grouping probably would expose the more vulnerable members when predators test the group. This would tend to place the burden of predation on older animals (Pimlott et al. 1969, Mech and Frenzel 1971, Mech and Karns 1977) that already have contributed genetically to the population, thereby increasing the chances of their own offspring to survive and breed. (5) Congregating would increase the ratio of deer to wolves in and near the yard, thus decreasing relative predation level through a sheer mathematical effect (Brock and Riffenburgh 1960). However this also has an important biological effect. Deer that summer in several wolf-pack territories may assemble in 1 yard. For example, deer that wintered in [one yard] summered around the edges of at least 4 wolf-pack territories. Although the wolves in whose territories the deer summer may also try to concentrate in or near the yard, their territorial nature tends to minimize the number of packs that can frequent a given yard. Competing packs will seek each other out and fight (Mech unpublished), often leading to the deaths of alpha animals [breeders] (Mech 1977d) [Mech 1977c]. (6) Individual herd members may have to spend less time alert and thus may be able to spend more time eating or ruminating than solitary individuals, as has been found in several other species (summarized in Hoogland 1979).

Besides their grouping in winter, and "spacing away" (i.e., spreading out into individual home ranges) in summer when fawns are most vulnerable, deer possess many other antipredator defenses. Newborn fawns, which are vulnerable if discovered by even small predators like foxes or fishers, are cryptically colored, thus matching the forest floor. They freeze in place motionless for their first week or so except when with their dam nursing. They also seem either to be odorless (which seems improbable) or to have their odor masked such that predators have great difficulty finding them. For example, dogs have been observed passing right by them (Severinghaus

and Cheatum 1956). When fawns reach 1–2 weeks of age, they no longer freeze in place but rather spring up when disturbed and bolt off at least fast enough that humans cannot catch them. They also bleat noisily, which no doubt serves to call the doe to their defense. Although no one seems to have observed a doe defending her young fawn against a wolf or wolves, a single deer has been known to stand off three wolves (Nelson and Mech 1994), and a doe was seen following a black bear that was carrying off her fawn (L. L. Rogers, pers. comm.). These observations suggest that does are very protective of their fawns. Does are also known to attempt to lure predators away from their fawns through a kind of "broken leg" act (Severinghaus and Cheatum 1956; Mech 1984).

Adult deer also possess many effective antiwolf traits, contrary to Stenlund (1955). Any deer hunter knows how keen deer vision, hearing, and scenting abilities are. These senses evolved in the face of wolves and other predators. And deer employ their senses well, usually by either running off as soon as they detect wolves or waiting until they confirm that wolves are approaching and then bolting off. Deer in dense cover are more wary and tend to flee sooner than those in more open cover, presumably because it is easier to detect an approaching enemy in the latter (LaGory 1986, 1987). As they flee, they "flag" their white tails. This flagging may demonstrate to wolves that the deer are fleet enough to escape, a potentially honest signal of condition that may inform wolves that it is useless to continue the chase (Caro et al. 1995).

When deer flee, they can speed away at 56 km/hr (Young and Goldman 1944; Severinghaus and Cheatum 1956), about the same speed as wolves (Stenlund 1955; Mech 1970). If wolves are close enough when the deer bolts away, they try to follow. However, most often, the deer bounds away, leaving the wolves behind. Usually the deer quickly senses that it is outrunning the wolves and then stops and watches its backtrail. If the wolves do persist, the deer bursts off again. It appears that the deer keeps trying to save its energy for when it really needs it.

Only rarely do wolves persist in chasing a deer for very long. However, if they do, as in one case of a single 2.5-year-old female wolf pursuing a deer for at least 20.8 km over a 2-hr period (Mech and Korb 1978), the deer was still able to stay ahead of the wolf. Unfortunately, the observation ended without the observers determining the outcome. The pursuing wolf and its pack were seen the next day feeding on a deer kill some 4.8 km from where the observation ended (Mech unpublished). Whether the kill was the same deer is unknown, but this incident is evidence of the deer's running endurance.

In some instances when a wolf or wolves do catch up to or confront a deer, then the deer employ other defenses. Bucks in autumn or early winter can kill wolves with their antlers (Nelson and Mech 1985), and adult deer at any time of year can kill with their hooves (Frijlink 1977; Nelson and Mech 1990). Thus some deer when confronted by wolves merely stand their ground and fight off the wolves (Mech 1984; Nelson and Mech 1994). These observations contrast to those of a sample of 16 white-tailed deer being approached by coyotes; those deer all fled (Lingle and Pellis 2002).

Another approach deer use to escape an attack by wolves is to flee into water (Joslin 1966; Pimlott et al. 1969; Mech 1970). Although a wolf has been observed killing a deer while swimming after it (account 47), fleeing to water is usually effective. The safety of large waterways probably explains why deer tend more to inhabit shores of lakes, points, and other edges of waterways during summer (Hoskinson and Mech 1976; Nelson and Mech 1999).

Besides the tendency for deer to live close to lakes, rivers, and other waterways for protection from wolves, another major way that white-tailed deer reduce their chances of predation by wolves, at least in Minnesota's Superior National Forest, is through their occupation of buffer zones between wolf-pack territories (Mech 1977a, 1977b). These areas around the periphery of wolf-pack territories constitute regions about 2–6 km wide where deer survive better than in the territory centers (Hoskinson and Mech 1976; Mech 1977a, 1977b; Rogers et al. 1980; Nelson and Mech 1981). The theory is that in these overlap zones, wolves have the best chance of encountering members of neighboring packs and thus are subject to altercation and death (Mech 1994a; Hayes 1995; Mech et al. 1998). Thus wolves spend less time in these buffer zones than in their territory center (Mech and Harper 2002). Each wolf visited a given deer home range along the edges of their territory an average of only about once per 20 d (Demma et al. 2007; Demma and Mech 2009b). Wolves also spent at least 1 hr/visit during five of eight recorded visits to these deer home ranges (Demma et al. 2007). Although comparable data for deer

deep inside a wolf-pack territory have not been available, wolf visits there no doubt would be much more frequent.

The deer summer spatial organization is based on a matriarchy in which female offspring remain near their mothers as they mature and begin breeding, forming a close-knit cluster of mothers and offspring home ranges (Nelson and Mech 1984, 1999). Apparently fawn survival is greater in these buffer-zone clusters than in wolf-pack territory centers, so these deer demes (Nelson and Mech 1987) persist there, whereas fawn survival is less in wolf-pack territory centers (Kunkel and Mech 1994), so deer demes are fewer, smaller, or nonexistent there. The end result is that deer tend to be found more along edges of wolf-pack territories than in centers. Evidence of similar wolf-deer dynamics within buffer zones was also found at the opposite end of Minnesota, where deer density was higher and wolf density lower (Fritts and Mech 1981). There is also theoretical evidence that buffer zones may be prey refuges (Lewis and Murray 1993) and that territorial stability in such zones requires interpack aggression by wolves (Taylor and Pekins 1991) as has been found (Mech 1994a). Thus, this type of deer refuge appears to be somewhat general. The extent to which it might apply to other prey species has not yet been tested.

Given the many antipredator traits, behavior, and other defenses that help deer survive in the face of wolf densities that can reach 182/1,000 km^2 (Mech and Tracy 2004) and pack sizes as high as 23 (Mech 2000), it is not surprising that deer do not often seem to allow wolf-predation risk to interfere with where they choose to forage (Kittle et al. 2008), contrary to Lima and Dill (1990) and Brown et al. (1999). Some deer even live as close as 800 m to active wolf homesites (Nelson and Mech 2000).

As deer in wolf range go about their daily lives, then, wolves are never far away and at any time may threaten any deer. The result is a constant tension between the two species, with each trying to survive by outdoing the other. This vital game is exemplified in each of the following hunting accounts based primarily on observations during winter from light, fixed-wing aircraft circling over the unfolding dramas in the Superior National Forest of northeastern Minnesota (unless otherwise indicated). Many of these observations resulted from Mech or his assistants homing in on radio-tagged wolves and finding them hunting deer. Those observations not made by Mech are attributed at the beginning of each to those who made them or reported them. The wolf identification numbers are given in each account, but the age and sex of the radioed wolf is given only where that animal is the only one involved in the attack. Where there are additional pack members, the identities of the accompanying wolves are unknown. Unless mentioned otherwise, the sex and age of the deer were unknown. The word "we," when used in the aerial observations, usually includes the pilot.

Hunting Accounts

1. December 28, 1953—Superior National Forest, Minnesota (Stenlund, 1955, 32)

R. Hodge and L. Nelson observed a large wolf trotting on the ice about 100 m away from a doe and fawn. The fawn ran. The wolf gave chase within 15 m of the doe, which remained standing. Within 100 m, the wolf knocked the fawn down, grabbed it near the right hip, and shook it vigorously. The fawn arose, and the wolf immediately knocked it down. The fawn's hind feet were stretched out behind, and the wolf grabbed at the front shoulder and neck. When the plane approached, the wolf disappeared. The fawn remained down awhile, arose, and walked away. It left no blood trail and did not appear to be hurt.

2. January 1964—Superior National Forest, Minnesota (Mech 1966b)

While I followed seven wolves, one strayed away for a few minutes, while the six others approached two deer lying 100 m away. Neither wolf nor deer was aware of the other. When the wolves were within 30 m of the deer, two other deer bounded by, pursued by the straying wolf. The two deer and their pursuer alerted the bedded deer and the wolves. Instantly, the wolves gave chase, focusing on one deer, getting no closer than 50 m, and within 1 min giving up. This pack advanced to within 100 m of two more bedded deer, which arose and fled before the wolves sensed them. When the wolves reached the beds, they tracked the deer for less than 1 min and abandoned the track.

In a similar observation, eight wolves approached within 30 m of a standing deer and stopped as the deer detected them. After deer and wolves stared at each other for 1–2 min, the deer bolted, and the wolves chased. Most got sidetracked when flushing another deer, but the lead

wolf continued after the first. The deer quickly outdistanced the wolf, which gave up within 250 m. Five pack members reassembled and rested. One started toward a third deer 150 m away. Both began running. The wolf lost ground and gave up within 125 m.

3. August 31, 1967—Superior National Forest, Minnesota (Mech 1970)

At 1315 hr, C. Ream and R. Ream watched from a canoe as a young, buck deer ran with a wolf 5 m behind. The deer continued into a lake swimming. The wolf, detecting the Reams, left but reappeared about 20 m farther along the lake, continued another 60 m, and disappeared.

4. March 6, 1962—Ontario, Canada (Pimlott et al. 1969)

At 1115 hr, J. A. Shannon saw six wolves around open water 100 × 40 m with a 6-yr-old deer swimming to the south end. When two wolves ran there, the deer swam north and stood in about 60 cm of water. During the next 1.5 hr, individual wolves swam toward the deer several times. The deer would then swim away. The wolf would follow but soon give up. In 3 hr and 33 min at least one wolf entered the water 27 times for an average of 43 sec (5–105 sec). During three times, two wolves swam at one time, and once three were together.

The next day, the deer was still there at 0900 hr. At 1109 hr, it stood in water 1 m from shore, and six wolves rested 9 m away. By next check, at 1530 hr, the deer was consumed, and the wolves gone.

5. January 26, 1967—Superior National Forest, Minnesota (L. D. Mech)

We were following eight wolves, when at 1615 hr they were about 200 m from two deer, one lying and one standing. When within about 30 m of the latter, the lead wolf stopped, and the others caught up but stopped 8 m behind. The wolves and deer remained still while staring at each other, 30 m apart, for 1–2 min.

The standing deer bolted, and almost simultaneously the wolves pursued. The deer headed toward the other deer, which had arisen when the wolves were about 150 m away. The lead wolf followed the deer's trail, while the others flushed the second deer. We continued following the first. It had no trouble in snowdrifts, but the drifts hindered the wolf. The wolf followed the deer for

200 m and then gave up after losing ground over 220–250 m. Three wolves had stopped in the second deer's trail. We did not see the remaining wolves or the second deer. These three wolves joined the first, and all rested. At 1625 hr, one wolf started toward a third deer about 150 m away. The wolf chased the deer about 125 m and gave up after losing ground. The deer continued running for at least 400 m.

6. February 25, 1967—Superior National Forest, Minnesota (R. Hodge)

At about 1130 hr, we saw four wolves heading east on the frozen Basswood River. When 60 m from a doe and fawn, just inland, the wolves bolted toward shore. They slowly worked toward the deer, paralleling the shore. When they were 20 m from the deer, it bounded off. The wolves bolted toward the deer but inland of them, never got any closer, and gave up within 1 min (15 m). The snow hindered the wolves.

7. March 10, 1968—Superior National Forest, Minnesota (R. Hodge)

At about 1500 hr, we spotted a wolf chasing a deer 75 m ahead. The deer was walking fast, and the wolf seemed to be pacing it. When the deer crossed the frozen narrows of a bay, she broke into a run, and the wolf did, too, and shortcut the deer. The deer bounded away with no trouble and soon stood on a cliff watching her backtrail. As the wolf approached, she ran back off the cliff. The wolf never climbed the cliff but cut across the narrows in front it and reached the deer's new trail, thus gaining considerable ground. Again, the deer outdistanced the wolf when running through snow. When the deer hesitated at the base of the cliff, the wolf dashed across the ice and hit the deer from behind. While the deer was down, the wolf grabbed it behind the head. A minute later the deer lay still.

8. February 27, 1969—Superior National Forest, Minnesota (L. D. Mech)

Wolf 1059's pack of five at 1610 hr came to within 100 m of two deer standing alertly in a shallow draw. When at least two wolves came within 100 m, the deer fled, and the wolves chased them. The deer were in belly-deep snow and hesitated slightly at each bound but still ran fast. We could only see one wolf well. It also had a hard time, and

after about 250 m (100 to the deer's original location and 150 after the deer), rested about 10 min. The deer ran only about 200 m more and stood alertly for at least 20 min. The wolves went on.

9. March 27, 1969—Saint Louis County, Minnesota (L. D. Mech)

At 1500 hr while following adult male Wolf 1051, we saw a deer running on top of crusted snow, then stand and watch its backtrail. After 1.5 min, 1051 was running along the same route. We saw the deer run 100 m from the wolf and double back, paralleling its original route. When the wolf got near the doubling-back point, he lay down for 5 min. The deer fled for 350 m, stopped, and for several minutes faced its backtrail. The wolf gave up.

At 1640 hr, Wolf 1051 was within 100 m of Highway 53 and paralleled it 150 m inland. Suddenly two deer fled across the highway, as did the wolf. Dense forest then obscured our view, but eventually the deer started southeast down a woods road, and the wolf was 50 m behind and gaining. The wolf got within 6 m, and the deer headed around in a circle, with the wolf closing in from the outside in the thick forest. Neither emerged for at least 15 min, so we left.

10. January 5, 1971—Superior National Forest, Minnesota (L. D. Mech)

At about 1030 hr we homed in on female Wolf 2204 (at least 2-yr-old nonbreeder) when we saw a deer that had just crossed a lake near its center. Then we spotted 2204 halfway across the lake, following the track and entering the woods where the deer had. In a cutover area we saw both a deer and 2204 about 200–300 m apart. The wolf moved deliberately at a half run, following one of many deer and moose trails. She followed about 150 m on one track, then backtracked toward the deer, which had run off again toward three other deer.

During the next 45 min, the wolf chased several deer and perhaps the same deer several times. Each time, the deer stood until the wolf got to within 150 m and fled very fast, with high bounds. Within 400 m the deer greatly outdistanced the wolf, which did not seem to be running her fastest. Then the deer stopped and watched their backtrails. The wolf gave up within a few hundred meters and started on another trail. Once the wolf came to within 100 m of a deer but was soon outdistanced.

Another time, a deer made a circle about 200 m in diameter and looked back toward the wolf. The deer then stood on a 150-m ridge and watched as the wolf passed some 15 m below. The deer then ran off, and the wolf took after the deer but again was outdistanced. The wolf returned to the center of most of the chasing and headed toward a small clump of conifers. A deer dashed out, with 2204 within a meter of it. The deer soon increased its lead to about 15 m in the next 150 m.

Then the deer stopped, and the wolf came to within 7 m. The deer rushed the wolf and hit her with both front feet, bulldozing the wolf. When the wolf arose, which she did immediately, the deer charged again and chased her about 7 m. Three times the deer charged the wolf, and twice the deer jumped right over her, kicking its hind feet back on landing on its front feet. Then the two reached a standoff with each circling slowly and cautiously.

At 1054 hr, the wolf curled up and rested while the deer stood facing her about 7 m away, and at 1210 hr when we returned, they were still in the same positions. At 1555 hr, we found both animals gone and no sign of a kill.

11. January 11, 1971—Superior National Forest, Minnesota (L. D. Mech)

We observed adult, male Wolf 1069 at 1430 hr near a deer for several minutes. The deer bolted, and the wolf came to within 7 m. However, the wolf fell far behind after 150 m, and gave up. Snow about 37 cm deep.

12. January 18, 1971—Superior National Forest, Minnesota (L. D. Mech)

At 1530 hr we watched as adult, male Wolf 1069 came on three deer, which fled when he was within 50 m. He chased one 100 m and gave up as the deer easily outran him. The snow was fluffy and about 30–45 cm deep.

13. March 8, 1971—Superior National Forest, Minnesota (L. D. Mech)

At 1755 hr we observed five wolves jump a deer but not get within more than 150 m of it. The deer ran about 200 m, stopped, and looked back for at least 5 min. The wolves had given up within a few seconds. The wolves were traveling on top of snow crust most of the time but occasionally fell through.

14. March 10, 1971—Superior National Forest, Minnesota (D. Hiser)

We spotted Wolves 2215 and 1075 and three others chasing four deer. The wolves seemed to give up after a few seconds but began the pursuit again. The lead wolf ran very fast, outdistancing his companions and closed quickly. The deer, together in a line, circled around a hill. The other wolves cut across and reached the deer simultaneously with the lead wolf. The last deer was struck and killed within less than 400 m, apparently by a shortcutting wolf. One or more wolves continued after the remaining deer and killed one about 200 m farther.

15. July 26, 1971—Superior National Forest, Minnesota (L. D. Mech)

At 1220 hr, we observed a deer bounding for 15 m, and a wolf nearby. The deer stopped, and the wolf approached within 15 m and stopped. The standing deer looked toward the wolf for 20 sec and bounded off. The wolf instantly bounded after it and came within 7 m when the two disappeared from our view.

16. December 20, 1971—Superior National Forest, Minnesota (L. D. Mech)

At 1305 hr, we radio tracked pup Wolf 2411 and three others approaching two deer 200 m away standing alert. At 1310 hr, the wolves jumped a deer 100 m away, which sprinted for 400 m. The wolves did not chase it and might not have seen it, although they pointed toward it. At 1315 hr, the wolves were running, a deer running 50 m ahead. The deer started walking alertly back toward the wolves, which had only run a few meters. One wolf got within 15 m before the deer ran. Then we saw another deer. The wolf chased one but gave up within 100 m after quickly losing ground. Another wolf chased a third deer but gave up quickly. At 1325 hr we left, but there was more chasing we could not observe because of trees. Snow was 16–30 cm of fluff.

17. December 20, 1971—Superior National Forest, Minnesota (L. D. Mech)

At 1435 hr we located breeding female Wolf 2407 and her mate and three deer 100 m ahead. The wolves passed 30 m from the deer, stopped and started toward them.

Two deer immediately fled at top speed, but the wolves did not follow. They headed toward the third deer standing 30 m from the others.

The lead wolf got within 15 m, and the deer fled, with the wolf in pursuit. The deer ran straight through a jack-pine plantation about 600 m and then veered 400 m. Several times it stopped when 200 m ahead and watched its backtrail for several seconds. As the wolf got within about 100 m, it fled again. Each time the initial burst showed the deer was refreshed.

Once, the deer curved around the base of a low hill 150 m across. The wolf tried cutting across but stopped after about 50 m, returned to the deer's trail, and sped on down it. As the wolf approached, the deer fled again in a horseshoe shape and back on its backtrail. The wolf followed 100–200 m behind. The second wolf, whose whereabouts had been unknown, came running along the trail of the first. This wolf then, and the deer, were heading toward each other but 30 m apart, with a row or two of conifers between them.

Meanwhile, the first wolf continued chasing the deer generally along its backtrail. We then lost the wolf and deer at 1457 hr, but it appeared that the wolf had given up. Snow depth was 15–30 cm and didn't hinder the deer and hindered the wolves only a little.

18. February 22, 1972—Superior National Forest, Minnesota (L. D. Mech)

At 1035 hr we spotted a gray wolf lying on the ice just west of open water, perhaps 3 × 3 m, and a black wolf standing at the east edge. A deer was treading water. The gray wolf joined the black, and the deer left the water and headed toward the wolves a meter away. The wolves grabbed the deer's front end, pulled it from the water, and bit its head, neck, and shoulders. Within 5 min it died, and the wolves fed (about 1045 hr).

19. February 22, 1972—Superior National Forest, Minnesota (L. D. Mech)

We aerially watched Wolf 2403 and two others chase four deer unsuccessfully. The deer were in two groups of two, but the chases were separate. The deer fled when the wolves were within 100 m, ran 150–250 m, stopped, and the wolves gave up. Snow light and fluffy, about 45 cm deep.

20. February 24, 1972—Superior National Forest, Minnesota (L. D. Mech)

At 1440 hr, Wolf 2415 and at least three others were chasing a deer over open ridges when we found them. The deer was 250 m ahead, and the wolves gave up within about 30 sec after we saw them.

21. February 24, 1972—Superior National Forest, Minnesota (L. D. Mech)

At 1630 hr, we watched a wolf chasing a deer east across Birch Lake. Both were running hard, the deer 300 m ahead. The deer ran onto a large island, then back off of it and to the lakeshore. When the wolf followed to the mainland, the deer cut back to the island, then to a nearby one and to a third, then back to the mainland. On reaching shore, it stopped out on the ice, for 30 sec. The wolf was just leaving the last island. At 600 m the wolf saw the deer and veered straight toward it. The deer ran back to shore, and the wolf followed. Neither was seen during the next 5 min, and the last we saw, the wolf was 100 m from the deer.

The next day, we checked the area again, and tracks indicated that a deer had come back to the lake and to the large island, then to the smaller, and had been killed just on the northwest side of the smaller island. Total minimum chase distance from where we first noticed it was 6 km.

22. February 15–25, 1972—Superior National Forest, Minnesota (L. D. Mech)

In the Prairie Portage area along the Minnesota border with Ontario between February 15 and 25, 1972, we had found several kills. On February 15, while tracking Wolf 2425, we found at least three other wolves near her. A fresh kill lay about 200 m inland. On February 17, a fresh kill lay on the bay west of the portage. The next day, we found nine wolves, a single wolf, and Wolf 2425 nearby. In the same area on February 22, we watched two wolves kill a deer there (account 18) and located two other fresh kills nearby. On February 24, we found another fresh kill, and while we circled, a wolf killed a deer beneath us, which we found just after the wolf killed it. We then watched the February 24 chase on Birch Lake (account 21).

Thus in 11 d, we located eight fresh deer kills near Prairie Portage. All were completely consumed. The 11 wolves, presumably all from the same pack, were all split up traveling across islands and on shore, and they were running deer onto the lake ice and near a treacherous river with thin ice and killing them there.

23. February 9, 1973—Superior National Forest, Minnesota (J. Renneberg)

At 1130 hr, we watched Wolf 2443, Wolf 2445, and two others begin chasing a deer. It appeared the deer had been surprised because the lead wolf was not more than 30 m behind. The chase continued along a ridge and down a short hill, with the wolf 15–20 m behind. While they descended the hill, two other wolves came to help 200 m behind. When at the bottom of the hill, the deer turned to face the oncoming wolf. The wolf all but launched itself at the deer as if to knock it down. But the deer stayed upright and held the wolf off. They circled around awhile before the wolf grabbed its head or neck. After more struggling the deer dropped, and one other wolf arrived to help finish it off. The chase took 4 min, and the total chase distance was 800 m.

24. February 15, 1973—Superior National Forest, Minnesota (L. D. Mech)

We watched Wolves 2407, 2247, 2489 as they began chasing a deer at 1615 hr. They never got closer than 100 m, and after a chase of 250 m, gave up. The deer then stood for at least 10 min about 200 m away.

25. February 22, 1973—Superior National Forest, Minnesota (L. D. Mech)

At 1400 hr, we found Wolf 2480 and four others. One bounded up and down as if trying to see something as we watched. A deer was standing 200 m away looking toward the wolves. They saw a second deer 100 m away and gave chase but gave up almost immediately. A third deer just watched the whole time.

26. April 17, 1973—Superior National Forest, Minnesota (J. Renneberg)

At 0725 hr, when I first located the Jackpine Pack, we saw a wolf chasing a deer into Pike Lake. They were 10–15 m apart. The wolf turned back to shore while the deer crossed to the other side and encountered some newly formed ice. It then made very poor headway. At 0737 hr, Wolves 2449, 2443, plus three others crossed the lake

narrows toward the tiring deer. When a wolf ran out on the ice and confronted the deer, it made no attempt to move away. It was almost as if the deer didn't know the wolves were there. The deer and wolf were a meter apart, and the deer immediately swam directly away. The wolves made no attempt to chase it or enter the water. They spread out along the shore and watched the deer. At 0750 hr, the deer began swimming in circles, having trouble keeping its head up and soon died. We left at 0753 hr. By 1020 hr the wolves had made no attempt to pull the deer out.

27. December, 13, 1973—Superior National Forest, Minnesota (J. Renneberg)

We located adult female Wolf 5071 at 1140 hr, 50 m behind an adult doe. The wolf never got closer than 45 m, and at times was 200 m behind. After chasing for 10 min, the wolf gave up.

28. December 4, 1974—Superior National Forest, Minnesota (L. D. Mech)

At 1140 hr, female pup, Wolf 5139, was seen chasing a deer near a fresh kill and two other deer. The deer found the other two, and all bounded away when the wolf was within 15 m. The wolf followed about 100 m, gave up, and returned to the kill.

29. December 31, 1974—Superior National Forest, Minnesota (J. Renneberg)

I located female Wolf 5176 (pup) chasing four deer, about 75 m behind them. She never got close.

30. December 10, 1975—Superior National Forest, Minnesota (J. Renneberg)

At 1605 hr, I saw Wolves 5196, 2248, and seven others chasing two deer. They gave up when the deer gained 150 m on them.

31. December 15, 1975–Superior National Forest, Minnesota (J. Renneberg, S. Knick)

At 1438 hr, we found Wolves 5417, 5419, 5429, and 5187 plus eight others, and saw two deer running. One to two minutes later, two wolves started trailing one of the deer. The deer stopped, and a moose nearby moved toward it. The deer started running again. The moose moved right in front of the wolves and made no attempt to run from them; it even moved toward them a bit. One of the wolves ran from it. When the moose saw the rest of the pack, it lumbered off. The wolves continued after the deer.

The deer had stood again until the wolves got to within 35 m, then ran off. The wolves ran toward the deer, and when they saw it, gave chase (75 m between). But the deer easily outdistanced them. The wolves continued another 400 m. Then the deer mixed in with four other deer, and straggling wolves began to catch up (1510 hr). We could not tell which deer was which. The wolves came within 20 m of one deer, but the deer easily outdistanced them. They continued after it and disappeared.

32. January 13, 1976—Superior National Forest, Minnesota (J. Renneberg)

At 1406 hr, we observed a wolf headed toward three deer. The wolf got within 20–25 m, but the deer then easily outran it. After a 75–100 m total chase, the wolf gave up.

33. January 28, 1976—Superior National Forest, Minnesota (J. Renneberg)

At 0940 hr, I located Wolves 2248 and 5196, plus at least four others. They came to within 30 m of two deer, but the deer easily outran the wolves. The wolves gave up after just covering the 30 m to where the deer started.

34. January 29, 1976—Superior National Forest, Minnesota (J. Renneberg)

At 0900 hr, I saw at least 2-yr-old female Wolf 353 chasing two deer but never getting closer than 75 m

35. December 19, 1976—Superior National Forest, Minnesota (S. Knick)

At 1445 hr, when we spotted Wolf 5429 and six others, they were chasing a buck deer running at top speed 250 m ahead. The deer ran about 300–400 m, stopped and looked back. When the wolves (always chasing in single file) got within 100 m, the deer ran again for 300–400 m and stopped. When the wolves got about 75–100 m away, the deer ran again, this time for a shorter distance. The wolves appeared to be tracking the deer. The deer always ran when the wolves got within 100 m and always kept the wolves upwind of it.

At 1509 hr we lost sight of the deer, and after 400 m

the wolves had lost the trail. Total chase distance was about 4 km, over 24 min. Wolves never got closer than 50 m.

36. December 28, 1976—Superior National Forest, Minnesota (S. Knick)

At 1115 hr while locating Wolves 2407, 5465, and 5469, plus two, we spotted two deer. Three wolves were sleeping about 75 m from them, and two were 40 m from them. The two wolves detected the deer and crouched behind some bushes. The deer was unaware of the wolves and proceeded another 10 m or so toward them. Then the deer stiffened up and became alert. They continued toward the wolves another 10 m, taking a few cautious steps, stopping, then some more. When the deer were about 20 m away, the chase began. The deer split up, and the two wolves pursued one of them. After about 150 m, the deer had a good lead, stopped, and looked back. When the wolves were 30 m away, the deer ran again, easily outdistancing them. The wolves pursued another 50 m and gave up. Total chase distance—200 m. Closest wolves got to deer—20 m.

37. January 1, 1977—Superior National Forest, Minnesota (S. Knick)

At 0940 hr, when we spotted Wolf 353 and two others, they were chasing two deer 50 m ahead. One deer stopped, but the others continued. When the lead wolf got within 30–35 m, the stopped deer fled at top speed, but the wolf gained to within 15–20 m. About 50 m later, the deer turned sharply left, but the wolf continued straight and lost the trail. At 0943 hr, the wolf stopped and joined the other two wolves. All three continued on while the deer were about 200 m away looking back. Total length of chase, 350–400 m, in 3 min. Closest distance between wolf and deer—15–20 m.

38. January 2, 1977—Superior National Forest, Minnesota (M. Korb)

At 1525 hr when we arrived, four wolves (Wolf 5401 and pack) were running full speed, two on a lake 30 m apart and two on a ridge north of the lake. The closest of three deer was about 100 m ahead. The other two deer traveled 200 m before stopping. The wolves gave up after a few hundred meters. The entire chase lasted 2–3 min from when we arrived until the wolves turned back.

39. January 24, 1977—Superior National Forest, Minnesota (M. Korb)

At 1135 hr when I arrived, five wolves (5407, 5448, and pack) were looking toward two deer, and the deer, downwind of the wolves, stood facing the wolves. The deer knew the wolves were present (30 m upwind) but could not see them. The wolves seemed aware of the deer but unable to pinpoint them. After several minutes, the deer bounded away. The lead wolf was the only wolf to give chase and did so for only 50 m and gave up. A knoll between the wolves and deer blocked visual contact between the two.

40. November 25, 1977—4.5 km Northwest of Isabella Lake, Superior National Forest, Minnesota (Mech and Korb 1978)

At 1210 hr, Wolves 2270, 5079, 5415, and a pup chased an adult doe while we watched. Wolf 5415, a 2.5-yr-old female, shortcut the deer and approached to within 40 m. The deer headed 800 m east, doubled back, and headed west. The remaining wolves abandoned pursuit, but 5415 persisted.

For the next 2 hr the deer ran west with 5415 in pursuit through varying cover and topography ranging from frozen lakes to spruce bogs and hardwood ridges. Once the wolf got to within 5 m of the deer, but at other times it fell about 800 behind. As the wolf lost ground, the deer usually slowed, bursting anew when the wolf gained. The deer was last seen about 1400 hr with the wolf 300 m behind, about 14 km from where the chase had begun. At 1420 hr, when the observation ended, Wolf 5415 was still following the deer's trail. This point was 20.8 km travel distance and an 18.1-km straight line from the starting point.

41. February 3, 1978—Superior National Forest, Minnesota (S. Knick)

At 1320 hr when we arrived, Wolf 5430 and seven others were sleeping in an open swamp. A deer was feeding 75 m away. By 1326 hr, the deer was within 50 m of the wolves. Neither deer nor wolves appeared to be aware of the other. At 1336 hr the deer froze, then charged the wolves but stopped 25 m away. The wolves stood staring at the deer and at each other about 20 sec; then one rushed after the deer, which turned and ran with high bounds through chest-deep snow. The wolf got within

10 m, but once on higher ground the deer easily outdistanced the wolf. The wolf gave up 25 m after reaching high ground. Total length of chase—100 m; closest wolf got to deer—10 m. Only one wolf really was in the chase. Another ran far behind.

42. November 27, 1978—Superior National Forest, Minnesota (M. Nelson)

At 1045 hr, we located Wolf 5931, 5465, another adult, and two pups traveling south 75 m in from a lake shore. Two deer 70 m east of the wolves were near open water and thin ice. The wolves climbed a ridge single file and stopped directly west of the deer. The two adult wolves split 25 m apart and continued south. The deer detected the wolves and walked fast south along the shore. Another deer directly in the path of the wolves stood facing them with ears forward. The wolves continued walking within 10 m of the deer before it ran. I don't think the wolves saw the deer until it ran. The deer ran south for 50 m and turned west. One wolf was 3–4 m behind it, and the other running parallel 25 m to the west. As the deer turned west, the second wolf did not seem to know the deer was crossing its path and never accelerated. The deer then broke onto an open ridge and increased its lead. The wolf suddenly gave up. The entire chase covered 100–150 m and lasted about 35–45 sec.

43. March 7, 1980—Superior National Forest, Minnesota (Nelson and Mech 1993)

Eight wolves of the Snowbank Lake Pack were aerially located traveling east, single file, through 43 cm of snow, which was a minimal hindrance to both wolves and deer. The pack abruptly reversed direction and ran just 1–2 m behind and alongside a bounding deer. The deer apparently had run into the wolves, probably chased by separated pack members and unaware of the wolves ahead. Within 1–2 sec, the deer had increased its lead to 10–20 m. Within another 200–300 m, only one wolf pursued the deer, which was pulling ahead. After the wolf gave up, the deer stopped and looked back several times before leaving.

44. December 22, 1980—Superior National Forest, Minnesota (M. Nelson)

At 1545 hr, Wolves 6037, 5926, and three others were traveling straight to a bedded adult buck. The wolves sensed the deer 75 m away, and the lead wolf sat looking at it. The deer remained bedded facing away and was apparently unalarmed. One wolf from behind the sitting wolf approached the deer at a slow walk. At 40 m from the deer, the wolf stopped as the deer stood and faced it for about 5 sec, and the deer bolted. The wolf chased but gave up after 75–100 m. The deer ran until 800 m away. The snow was 30 cm deep and did not hinder the deer.

45. February 19, 1981—Superior National Forest, Minnesota (M. Nelson)

Wolf 6089 and six associates split up while chasing a deer. The deer was running 75–100 m ahead of the lead wolf. One wolf closed the distance remarkably fast running downhill to within 25 m of the deer. The wolf must have been slowed by some downed trees because the deer surged 500 m ahead, and the chase ended. The chase distance was 800 m, for 1–2 min. Snow depth was 20 cm and soft.

46. February 19, 1982—Superior National Forest, Minnesota (M. Nelson)

We located Wolves 6123 and 6113 and pack mates. Three wolves were 10 m behind a buck deer running slowly onto a lake. One closed the distance and grabbed the deer's hind legs. Another approached from the side. The deer initially stood while the wolves tore at it for at least 1 min, then collapsed 100 m from shore. Snow depth on the lake was 40 cm, and 65–80 cm in the woods.

47. November 15, 1982—Superior National Forest, Minnesota (Nelson and Mech 1984)

At 1045 hr, we observed a radio-collared wolf with six others walking on shelf ice along the southwest side of Thomas Lake (1.6 km by 4.0 km). The rest of the lake, except for smaller bays, was open. An adult doe was swimming toward the opposite shore. The wolves were traveling on land toward the same shore. About 100 m from shore the deer turned back to the open lake.

We returned 3.5 hr later, and the deer was still swimming there; the wolves lay along the nearest shore. As the doe swam within 75 m of shore, a wolf jumped in and swam 25 m toward her before turning back. The doe swam parallel to the shore toward an island 400 m away. The wolves headed along the shore toward the island, which ice connected to the mainland. The doe reached the island first but remained in the shallows. As the wolves approached to within 100 m, the doe swam away. One swam after her but only came within 50 m before turning back.

An hour later, the deer was swimming toward the same island, and the wolves moved to intercept her. Now she swam more slowly and rested in shallows near a 5-m cliff that enabled the wolves to approach undetected. The wolves were 25 m away before she bounded toward the open lake. A wolf swam after her to within 2–3 m. When about 75 m from shore, the doe turned slightly, and the wolf caught her. The wolf swam alongside the doe and bit her neck and head. In the ensuing minutes, single wolves swam out twice toward the pair, but one returned when only 10 m from shore. For about a minute, two wolves attacked the doe, but then one returned to shore. Only one wolf was attacking as the deer died. The wolf towed the doe to shore, and all fed on it. The doe had been in the water at least 4.7 hr since we first observed her.

48. January 12, 1984—Superior National Forest, Minnesota (M. Nelson)

At 1337 hr, I located Wolf 6556 and six pack mates on Big Island in Eagle Nest Lake no. 3. After a few minutes, a deer ran onto the lake in full stride with a wolf 100 m behind. The deer slipped slightly halfway to another island. Three wolves were chasing, at a full run. As the deer was 75 m from the island, the lead wolf closed to within 50 m, and the wolves caught the deer 25 m onto the island. The chase lasted 2–3 min. (There were 45–60 cm of snow on island; also many blown-down trees; 5–15 cm snow on lake.)

49. March 1, 1984—8 km East of Tower, Minnesota (M. Nelson)

I found Wolf 6556 and seven others at 1511 hr and saw two of them chase a deer down a frozen creek. The wolves were 50 m from the deer when they suddenly stopped. The deer stopped then also. Wolves and deer looked at each other for 10 sec, and the wolves walked back toward the others, now traveling away. The deer walked toward the wolves for 50 m but stopped as the wolves entered heavy timber. Some 500 m beyond, the wolves came to a freshly killed fawn, and two bit its neck and shook the deer. Apparently the wolves had killed the deer while I watched the other, probably its mother.

50. May 17, 1984—Superior National Forest, Minnesota (E. Proetz)

Ed Proetz, Superior National Forest, reported that in June or July 1983 he observed from the ground a deer on a point, and 15 m behind was a wolf. The deer went to a small island, just 2 m from the point, and the wolf jumped at the deer's back. The deer just crumpled from momentum. Both splashed in the water, and the wolf ran back onto the point. Probably it saw Proetz and canoe 45 m away. The deer then swam from island to island and crossed to the other side.

51. June 12, 1985—Superior National Forest, Minnesota (M. Nelson)

At 1308 hr we located breeding, female Wolf 6494 (at least 4 yr old) chasing radioed, female Deer 6730 in a wet, sedge meadow for 100 m. It stopped as the deer entered a small stand of timber. The wolf had gotten within 25 m of the deer, lay down for a minute, and left. Deer 6730 stood 100 m beyond where the chase ended and looked back toward the wolf.

52. December 16, 1986—Superior National Forest, Minnesota (M. Nelson)

About noon I aerially located Wolves 6895, 1831, and 6767. Adult male Wolf 6895 started to chase three deer that had surprised the bedded wolves. The wolf came within 30–50 m of the deer; one deer split off, and 6895 pursued it. The deer ran within 10 m of another bedded deer. The wolf had slowed to a fast walk or slow run as it followed the first deer. Wolf 6895 also came within 10 m of the bedded deer but had his nose down and continued after the first. The bedded deer remained bedded as the wolf passed. After 100 m, the wolf lost the deer track and spent at least 5 min circling the ridge and crossing back and forth, then gave up.

53. November 1, 1986—Superior National Forest, Minnesota (M. Nelson)

At 1230 hr, I watched breeding female Wolf 6027 (at least 6 yr old) trying to pull down an adult buck on glare ice of Bogberry Lake. There were only 3–5 cm of snow, and the deer was slipping as it tried to run. Its rear legs slid apart,

and it fell and rested 1–2 min with the wolf also resting at its rear. The wolf lay completely on its side three to four times. When the wolf tried to move to the deer's front, the deer pointed his large antlers toward the wolf. As the deer stood up, the wolf pulled at its hind legs. Once the wolf rested with its teeth in the deer as the deer rested and looked at the wolf. Both rose and fell four to five times in 8 min. The struggle had proceeded for half the length of the lake and had almost reached shore. When 50 m from shore, the deer was exhausted. Some 20 min later the deer was dead, and two wolves were there.

54. December 28, 1986—Superior National Forest, Minnesota (L. D. Mech)

Homing on 2.5-yr-old female, Wolf 413, I saw two wolves traveling toward two bedded deer 150 m away (1520 hr). When the lead wolf was 100 m away, the deer fled. The wolf chased for 50 m and gave up.

We then saw Wolf 413 chasing another deer 100 m ahead. The deer gained easily, and after 200–250 m, it stopped and watched its backtrail. The wolf got within 100 m, and the deer fled again. The wolf continued chasing the deer, but we could no longer see the deer. The wolf stopped, sniffed the ground, and jumped two deer 25 m away and chased one. The deer pulled 75 m ahead, and the wolf chased it for 1.6 km. At times, three deer were running ahead of the wolf. Once the wolf ran within 35 m of a bedded deer, which remained bedded. After at least 12 min we lost the wolf still running. Snow was 33 cm deep and fluffy and no hindrance to wolf or deer.

55. January 18, 1988—Superior National Forest, Minnesota (M. Nelson)

At 1338 hr, I located Wolf 75 and saw five wolves circling the area as if hunting. One large and two smaller ones stopped on the ice of a river, and a deer ran across the river. All three wolves pursued and got to within 100–200 m. The deer entered some open water shallow enough to stand in. After a minute, the large wolf approached the deer in the water. As the wolf reached the water, the deer charged the wolf with head down, and the wolf fled. The deer stopped, and the wolf charged the deer and grabbed it by the neck, head, or ears. The deer dragged the wolf 10–15 m toward the water. At the edge of the ice, the deer fell to its side and struggled to get up. It then flopped into the water, and the wolf let go. Meanwhile, the other two wolves crossed and joined the attacking wolf. All three stood and stared at the deer standing in the water. They started to leave when a fourth wolf approached. The original three wolves kept on going, but the fourth stopped and watched the deer for 1–2 min. It then left. Thirty minutes later the wolves (now 11–12) were 800 m from the scene. The deer was gone, and I saw no blood. However, 3 d later two ravens were eating something there on thin ice, and there were many wolf tracks, so the wolves might have killed the deer.

56. February 25, 1988—Superior National Forest, Minnesota (M. Nelson)

At 1103 hr, while homing on Wolf 6439, we observed three bedded wolves, which started moving. Four deer were 500 m from the wolves, looking toward them. The deer then left. The next thing I saw was a wolf killing a deer 200–300 m from where the deer were first seen. Snow was 40 cm deep.

57. February 21, 1989—Superior National Forest, Minnesota (M. Nelson)

At 1309 hr, I saw a male fawn running hard, and 2–3 min later, male Wolf 135 was chasing it. The deer was 800 m ahead. A bounding deer track led to the deer out on Birch Lake. Meanwhile 135 followed the bound marks and tracks on the lake. The deer had headed onto a small island and then left it. The wolf saw the deer and ran toward it, but the deer continued walking unaware of the wolf. After 100 m, the deer ran back to the island with the wolf following by 50–75 m. Neither animal left the island, and the deer was killed there. The chase lasted at least 30 min, over 2.4 km.

58. December 6, 1990—Superior National Forest, Minnesota (M. Nelson)

I located Wolf 105 and one wolf moving fast. A deer 200 m ahead fled, and 105 chased but only got within 100–150 m. The chase lasted 30 min.

59. March 12, 1993—Superior National Forest, Minnesota (Nelson and Mech 1994)

At 1020 hr, we saw a large adult wolf (sitting), male pup Wolf 500 (sitting), and a third wolf (standing) 25 m from a large deer, standing and staring at the wolves. One wolf approached to within 5 m and circled slowly to the deer's opposite side. The other two wolves slowly approached to within 5 m of the deer, more or less equidistant from each other and the other wolf. The deer remained motionless. After 5 min of staring at the deer, the wolves left, and 15 min later were 1 km away. The deer remained motionless for that time.

60. Jan, 23, 1998—Superior National Forest, Minnesota (M. Nelson)

At 1032 hr, I radiolocated the Pike Lake Pack (six wolves) traveling east toward four deer, 300–400 m ahead, alert and looking toward the wolves. The wolves split into two groups and approached the deer on parallel routes. At 100 m away, two deer ran south. The wolves continued walking east, then one wolf ran fast east 300 m ahead of the rest, which then stopped and faced north. The other two deer were running east as well, 100 m ahead. One deer stopped to look back at the wolf, which continued fast in the track of the deer but out of sight of it. That wolf stopped, and the others caught up. After a short pause, one wolf again started on the deer trail for another 300–400 m. The deer, after slowing to a walk and looking back, again accelerated. This time the wolves did not follow.

61. August 20, 1999—Lake Vermilion, Minnesota (Carlson-Voiles, 2012)

Carlson-Voiles watched from her cabin on the lakeshore as a deer and then a wolf entered Lake Vermilion. Following are excerpts from her description of the encounter:

> The tall reddish ears of the deer were now perhaps 9 to 15 meters ahead of the furry ears of the wolf. . . . The wolf was gaining, moving now into the wake of ripples from the deer. . . . With a sudden burst of speed from the wolf, the two heads were now side by side. That's when the lake exploded. There was a thrashing of white water, a chopping of long legs, sharp hooves, a loud resonant bleating from the deer. It seemed to go on and on. But, in truth, it was probably just seconds before I saw the animals separate, deer pulling ahead. This time the deer's tail was about 1.8 to 2.4 meters ahead of the wolf's nose. But the wolf swam on. Again it gained on the deer . . . and again the water exploded with bleating and thrashing, waves now moving away from the disturbance in a widening circle. . . . But the wolf was beaten. The two

heads continued swimming, but this time the wolf swam straight back toward our wild shore while the deer continued toward the island.

62. June 2, 2006—Yellowstone National Park, Wyoming (M. Metz)

At 0753 hr, breeding male Wolf 295M arose as he saw a mule deer buck a few hundred meters from the Swan Lake Pack den and stalked toward the deer. Within a minute, he trotted toward the deer and disappeared. At 0756 hr, he reappeared chasing the deer, only in view for 30 sec. Ten minutes later, as the deer entered a river, we saw that it had an open wound on its rump. For 30 min, the deer moved between the shallower north side and deeper south side of the river, depending on which side 295M was on or moving toward. Wolf 295M also occasionally swam toward the deer. Once 295M grabbed the deer's hind end, but the deer quickly escaped. At 0838 hr, the deer was tucked against a steep bank, and 295M began crossing toward the deer. The wolf briefly grabbed the deer, and then quickly emerged from the river and shadowed the deer from the bank. At 0839 hr, the deer struggled in the water and collapsed at the river's edge while 295M watched. Wolf 295M immediately crossed the river while the deer was barely standing on its front legs. At 0840 hr, as the deer lay down, 295M pulled it onto the bank. The wolf avoided the front of the deer, and bit only its hind end, along its side near the spinal column, and at the back of its neck for the next 20 min while the deer occasionally attempted to get up or kick. Following the deer's death at 0902 hr, 295M rested for a few minutes before feeding.

63. July 24, 2014—Elbow Lake (Saint Louis County), Minnesota (C. Kirschner)

"We got up early to go fishing on the north arm of Elbow Lake and it was a very foggy morning. We were going pretty slow up the north arm at 6:15 A.M. and thought we saw something in the water from a distance. As we got closer, we saw two animals in the water, and realized it was a wolf chasing a fawn from one shore to the other. The wolf followed the fawn in the water and when the fawn got near the shore, the wolf jumped up on the fawn's back and bit it (I believe in the back of the neck). I finally found my phone and started filming right after this happened. The fawn was trapped in the water near shore

(stunned or partially paralyzed?) and the wolf made several attempts to finish the job and drag the fawn out of the water" ("Wolf Kills Fawn after Chasing across Elbow Lake—Cook, MN," filmed by C. Kirschner, July 24, 2014, https://www.youtube.com/watch?v=r7dDrrWu8Ag).

Additional accounts of wolves hunting deer, including diagrams of the chases can be found in *The World of the Wolf* (Rutter and Pimlott 1968).

Conclusion

It should be clear from the above observations both that deer possess many effective antiwolf strategies and characteristics and that wolves occasionally find ways of thwarting them. As indicated above, two main factors constitute the key to the interplay of these conflicting systems: (1) vulnerabilities of a small proportion of deer and (2) persistence by wolves in detecting those vulnerabilities. As with elk and other species, when wolves do detect vulnerabilities, whether actually perceiving the weakness (www.press.uchicago.edu/sites/wolves, video 7) or just by the nature of the interaction (Mech and Peterson 2003), their persistence pays off. This persistence takes the form of extensive travel mentioned earlier and enduring many unsuccessful deer encounters.

With wolf-pack territories averaging from 116 to 334 km^2, where the main prey of wolves is white-tailed deer (Fuller et al. 2003) and lowest deer densities are about 0.3–0.7 per km^2 (Nelson and Mech 1986a), generally 100–200 deer or more would be available for each pack of wolves during spring, summer, fall, and part of winter, depending on snow conditions. Packs constantly scan the deer for vulnerabilities.

The wolf's hunting strategies thus must vary to match the deer's lifestyle at any particular time of year. During May through October or November, for example, when does are back on their summer ranges and have borne fawns, wolves foray out from the pack's den of pups and often travel singly (Demma et al. 2007). They rarely kill adult deer at this time but concentrate on fawns (Nelson and Mech 1986b). To do so they must search a given doe's home range to find the fawn and then contend with the defensiveness of the doe when the fawn is quite young, and then, as a fawn gets older, wolves must chase and catch it.

Although no one has described wolves hunting deer fawns during summer, we have learned a few bits and

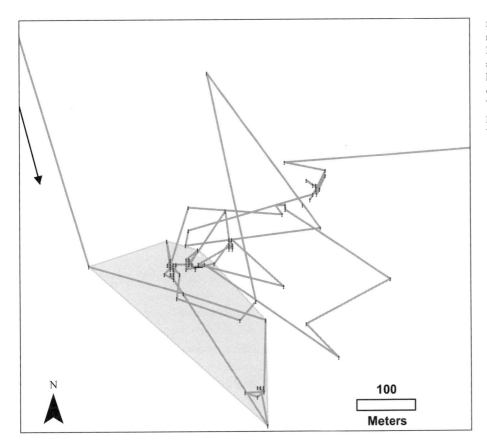

FIGURE 1.2. Locations of a radio-collared wolf in the Superior National Forest of Minnesota over a 21.8-hr period while attacking, killing, and eating a white-tailed deer fawn (Demma et al. 2007). The shaded area represents the presumed area over which the wolf chased the fawn.

N

100

Meters

pieces of information that yield insight into the process. We know that in at least a few cases wolves that kill fawns spend many hours and much travel associated with fawns they kill. In one instance when a fawn carcass had not yet been consumed, the wolf had already spent 11.5 hr and traveled a minimum of 1,306 m in the immediate vicinity (Demma et al. 2007). In two other cases, a wolf spent 21.8 hr and 20.1 hr and traveled minimum distances of 2,998 m and 1,503 m, respectively, in the immediate vicinity of the kill. These cases imply that it may take a wolf or wolves several hours of contending with a doe and chasing a fawn before success. The actual periodic locations of wolves during these interactions provide additional support for this conclusion (fig. 1.2).

Other information gleaned so far about wolves hunting fawns in summer includes a preliminary estimate that in the Superior National Forest at least one wolf visits each doe's summer range about every 3–5 d or about 18–30 times each summer (Demma et al. 2007). There is also evidence that wolves tend to visit summer ranges of older does (which usually produce multiple fawns) more often

than those of younger does (which tend to produce single fawns [Demma and Mech 2009b]). Furthermore, the fawns that wolves end up killing tend to be lighter weight (Kunkel and Mech 1994; but cf. Carstensen et al. 2009) and have abnormal blood values indicative of poor condition (Kunkel and Mech 1994; Carstensen et al. 2009). A preliminary estimate of the success rate of wolves hunting fawns during summer is 25% (Demma et al. 2007).

In autumn, wolves begin killing adult deer as well as fawns, and this tendency continues throughout winter and into April (Nelson and Mech 1986b). Usually by October or November, the surviving wolf pups have joined the adults on their nomadic journey throughout the pack territory as they search for vulnerable deer. The wolves seem to have little trouble detecting deer even in heavy cover, and that cover may be important in helping wolves conceal their approach (Kunkel and Pletscher 2001). From late autumn to early winter when most white-tailed deer tend to migrate from their summer ranges to winter yards (see the introductory material above), they become especially vulnerable to wolf predation. Nothing

TABLE 1.1.　Conditions that predispose white-tailed deer to wolf predation

Condition	References
Males in rut	Nelson and Mech 1986b
Older deer	Pimlott et al. 1969; Mech and Frenzel 1971; Kolenosky 1972; Mech and Karns 1977; Fritts and Mech 1981; Boyd et al. 1994; Kunkel et al. 1999; DelGiudice et al. 2002; DelGiudice et al. 2006 (but see Stenlund [1955] and reply by Mech [1970] reply; also see Potvin and Jolicouer 1988)[1]
Hoof injuries, abnormalities	Mech and Frenzel 1971
Deviant blood values	Seal et al. 1978; Kunkel and Mech 1994; Carstensen et al. 2009[2]
Poor nutritional condition	Mech and Frenzel 1971; Mech 2007a; DelGiudice 1998
Poor parental/grandparental condition	Mech and Karns 1977; Mech et al. 1991
Offspring of younger parents	Mech and McRoberts 1990
Low-weight fawns	Kunkel and Mech 1994
Migratory deer	Nelson and Mech 1991

Note: Table modified and updated from Mech and Boitani (2003).

[1]The authors concluded that with low deer density wolves did not tend to take old deer disproportionately. However their sample was small, and they did not consider deer in the old class until 7.5 yr old.

[2]Most of the fawn mortality was caused by bears, although some was caused by wolves.

is known about the cause of this increased vulnerability, but the deer mortality rate to wolf predation during fall migration was 16–107 times higher than for deer already yarded (Nelson and Mech 1991). Female deer that did not yard during December–March also suffered a higher mortality rate than those that did.

During much of winter, snow conditions become of critical importance in several ways. Not only does snow depth and consistency affect escape abilities of deer when wolves are chasing them, but it also greatly influences their condition (Verme 1963; Mech et al. 1971). Deep snow hinders deer movement and ability to feed and maintain their condition both directly and indirectly, as discussed earlier. Thus the nutritional condition of deer varies markedly among individuals and throughout the winter. Continued testing of each deer by wolves eventually exposes vulnerable animals, and the wolves feed. At the same time, deep snow often hinders wolves during their chases.

Besides general nutritional condition, many other factors can render deer vulnerable to wolves (table 1.1), and these have been discussed in detail elsewhere (Mech 1970; Mech et al. 1998; Mech and Peterson 2003). Suffice it to say here that a wide variety of data—everything from abnormal blood values, through old age, physical abnormalities, and even being offspring of grandmothers that experienced poor nutrition while carrying a given deer's mother—have been implicated in predisposing deer to wolf predation. In addition, we find no reason to question the 40-yr-old statement that, "un-

fortunately, even if human beings observed many acts of predation, we probably could not recognize most of the circumstantial, behavioral, and physical conditions that might have made the prey vulnerable. About the only conditions that might be recognized are a few physical ones that are evident from the remains of kills" (Mech 1970, 248).

Thus for most of the year, the wolf pack travels throughout its territory covering as much as 25 km per day at a rate of about 8 km/hour (Mech 1966a, 1994b), sifting the air for the scent of prey. When the wolves detect deer, they usually proceed slowly and deliberately, ever on the alert. They then tend to advance until the deer suddenly bolts and then spring after it. If the deer's initial burst and the circumstances of the encounter give the deer a head start, which is the usual situation, the deer quickly escapes, and the wolves give up. These frequent escapes by deer being chased by wolves belie the popular notion of the highly efficient hunter. Nevertheless, under certain circumstances as yet not well known, the wolves can chase the deer and catch up to it. Once they do, they lunge at it wherever they can get a good hold and often knock it right down.

Occasionally a deer stands its ground (account 10) and can hold off as many as three wolves (account 59). However, in most cases if a deer runs and the wolves catch it, the wolves have little trouble in bringing it down and dispatching it. In an assessment of some of the above observations, wolves closed to within 5 m of 16 deer and killed 12 of them (Nelson and Mech 1993). Because so many of the deer that wolves kill this way possess various vulnerabilities, it is reasonable to conclude that the wolves somehow detect the vulnerabilities, as is apparent in www.press.uchicago.edu/sites/wolves, video 7, and that is why they press the chase and catch the animal rather than quickly giving up.

The net result of the combination of deer vulnerabilities and wolf hunting persistence yields an unending series of wolf-deer interactions like those described above but, ultimately, relatively few successes. Hunting success rates differ with varying circumstances, and any calculated success rates are necessarily approximate (Mech and Peterson 2003). Nevertheless it is of interest to note how low they are.

Based on most of the hunting accounts described above, the estimated success rate of wolf chases of deer in northeastern Minnesota was 20% (Nelson and Mech 1993). However, if one bases hunting success on total number of deer the wolves interacted with, the success rate was 14 of 131 or 11%. This rate is not significantly different ($p = 0.25$; $\chi^2 = 1.66$; df = 1) from the 6% success of wolves attacking moose on Isle Royale (Mech 1966a combined with Peterson 1977). One should note the much higher estimated success rate of wolves hunting deer in Ontario, based on ground tracking in snow: 16 successes of 35 hunts, or 46% (Kolenosky 1972). However, the study's author warned that this rate may have been biased high because it is easier to detect from snow tracking a pack's success than its failures (Kolenosky 1972). As noted above, success rates would necessarily vary with snow conditions, season, and various other circumstances, but wolves often fail to overcome the antipredator defenses of deer. The bottom line, however, is that under all but the most extreme conditions (Mech and Karns 1977; Nelson and Mech 2006), both deer antiwolf behavior and wolf hunting behavior mesh well enough that both species survive.

2

Moose

THE MOOSE IS one of the wolf's most formidable prey. Standing about 2 m tall at the shoulder, bull moose weigh 380–720 kg and cows, 270–360 kg. The species' current range includes most of Canada south of the Arctic; Alaska; the northern parts of most of the northern United States, including Isle Royale National Park, and extending south in the Rocky Mountains into Colorado; and northern Europe and Russia, south of the Arctic. The only detailed discussions of wolf behavior hunting moose are by Peterson (1955) and Mech (1966a, 1970).

Wolves inhabit most of moose range, so at least in North America moose are one of the commonest prey of wolves. Like other prey, moose possess several modes of defense. They are generally solitary but sometimes live in groups of up to 12, especially in areas with little cover (Peek et al. 1974). They tend to seek conifer cover (Mech 1966a; Stephens and Peterson 1984; Kunkel and Pletscher 2000) and space away from wolf travel routes (Kunkel and Pletscher 2000). The farther moose are from cover, the more time they spend in an alert or alarm state (Molvar and Bowyer 1994). Moose can detect approaching wolves up to 400 m away (Mech 1966a) and possibly farther, and if they do, they often speed off, thus possibly avoiding a wolf encounter.

When wolves do find moose, these largest members of the deer family can call on their impressive physical abilities. Besides their sheer size and strength, moose can depend on their stamina to run and plow through deep snow and on their outright endurance during an attack, which sometimes lasts days. Long, coarse hair covering thick, tough hide helps armor moose against wolf fangs, while the moose aim their hard, massive hooves at the attacking wolves. In autumn, massive antlers probably help bulls fend off wolves. Although no one seems to have actually observed antler use, every other wolf prey

with antlers uses their antlers against wolves. Cows with calves (calves being more vulnerable) seek refuge in denser cover (Peek 1971; Peterson 1977) and away from other moose (Thompson et al. 1981). And if all else fails, moose readily take to water to stand off wolves (Cowan 1947; Peterson 1955, Mech 1966a).

Given the size and strength of adult moose, it is not surprising that wolves often target calves (Mech 1966a; Peterson 1977). Calves, usually born in May or June, weigh about 14 kg at birth, growing to perhaps 140 kg in late autumn, still much lighter than adults and much less powerful. Thus cow moose tend to select the safest areas in which to bear their calves, such places as islands, dense hiding cover, and high elevations (table 2.1). The problem that wolves face in trying to secure calves, of course, is the cow. And moose cows are extremely protective of calves (see e.g., "Pack of Wolves Attack Moose and Her Baby (Long Fight)," Video Malayalam, uploaded February 13, 2012, http://www.youtube.com/watch?v=ob3py1xa8p0), even threatening humans who happen to encounter the pair (Mech 1966a). Probably the most protective cow moose ever recorded was one that defended her two dead calves (9 mo old) against wolves for 8 d during winter in Denali National Park (Mech et al. 1998).

Despite this extreme level of calf protection, wolves sometimes prevail because in the inevitable maneuvering around in the brush that cow and calf have to do when attacked by more than a single wolf, the cow cannot always fully protect a calf, especially young ones in summer when most vulnerable (Mech 1977d, 532–33). Unfortunately, there are only a few descriptions of wolf attacks on moose calves in summer, but it is clear from these accounts that the predominant feature of such interactions is that the cow and calf try to stay close together while the wolf or wolves try to separate them and grab the calf.

TABLE 2.1. Antipredator moose calving locations

Location	References
Dense hiding cover	Altmann, 1958, 1963; Stringham 1974; Leptich and Gilbert 1986; Costain 1989; Langley and Pletscher 1994
Islands	Seton 1927; Clarke 1936; Peterson 1955; Stephens and Peterson 1984
Proximity to water[1]	Bailey and Bangs 1980
Highlands	Wilton and Garner 1991
Islands of cover in clearcuts	Cederlund et al. 1987

[1]To allow escape into water.

TABLE 2.2. Percentage of moose-calf mortality to wolves during summer

Location	N	% killed by wolves	Reference
Kenai Peninsula, Alaska	47	6	Franzmann et al. 1980
Yukon Flats, Alaska	80	1	Bertram and Vivion 2002
Nowitna and Koyukuk, Alaska	151	9	Osborne et al. 1991
Southwest Yukon Territory	119	25	Larsen et al. 1989[1]
South-central Alaska	120	3	Ballard et al. 1981
East-central Alaska and Western Yukon	33	12–15	Gasaway et al. 1992

[1]Year around.

The cow remains highly defensive of the calf. However, several studies have found that wolves take only 1%–15% of moose calves in summer, and one study learned that wolves killed only 25% of calves during their entire first year (table 2.2). These findings demonstrate the effectiveness of moose-calf defenses.

When wolves confront a cow and calf in winter, when calves are older, the calf still tries to stay in front of the cow. This tactic keeps the cow in an ideal position to protect the calf's rump, the usual initial point of a wolf attack. Each time a wolf tries to jump at the calf's rump, the cow strikes out at the wolf with its front hooves. Such defense is very effective and often keeps wolves from getting even a quick bite of the calf. After calves are a few months old, their own front hooves grow increasingly effective at any frontal assault from wolves.

Many moose merely stand their ground to ward off wolves. Mech (1966a) learned that dramatically early in his career when he watched a pack of 15 wolves attempt to attack a moose on Isle Royale. Having only begun to learn of the details of wolf-moose interactions, Mech assumed that such a large pack could easily kill a moose they had right in front of them (fig. 2.1). Within 5 min of confronting the moose, however, the wolves gave up their attempt and continued on their way (account 15).

This incident was only one of many examples of how formidable moose defenses usually are. Standing their ground and defying wolves is not the only way moose defend themselves. Sometimes they flee as the wolves approach and merely outrun or outlast the wolves. Or the moose flee first, and if their attackers catch up to them, then they stand and fight. On Isle Royale, 63% of 180 moose attacked by a pack of up to 15 wolves outran the wolves, 13% ran but then stood their ground, and 24% immediately stood their ground (Mech 1966a; Peterson 1977). Rarely do wolves kill a moose that stands and fends them off unless it is already wounded. Rather, it is almost always a running moose that wolves attack, although wolves do not attack all moose that run (fig. 2.2). As wolves chase the moose, they try to attack its rump, but the moose usually counters by kicking up its hind hooves and lashing out with its front ones. An attacked moose sometimes stops and fights its attackers. One solid blow with a hoof could seriously injure or kill a wolf, and several cases of moose injuring or killing wolves have been documented (MacFarlane 1905; Stanwell-Fletcher 1942; Mech and Nelson 1990; Weaver et al. 1992.)

Hunting Accounts

All the following hunting accounts except two are aerial observations from fixed-wing aircraft, except where noted. The sex of moose is unknown unless indicated otherwise. Distance was visually approximated.

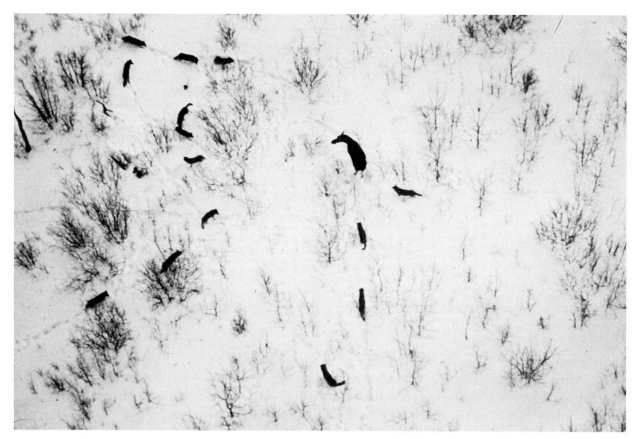

FIGURE 2.1. This moose stood its ground for 5 min, then the wolves gave up (account 15; Mech 1966a).

1. February 24, 1959—Isle Royale, Michigan (Mech 1966a)

Ten wolves from a pack of 15 were traveling southwest on Lake Richie at 1300 hr when they pointed upwind for a few seconds. Then they started inland, single file, upwind toward two adult moose feeding 400 m away. When the pack was within 200 m, the moose fled, one toward Lake Richie, the other away. The wolves chased the latter through deep snow, but soon all but the first gave up. It got within 7–14 m of the moose but then stopped. The others rested 100 m behind.

2. February 24, 1959—Isle Royale, Michigan (Mech 1966a)

At 1800 hr, 10 of the 15 wolves were traveling along the shore of Siskiwit Lake 1.6 km ahead of the others. They stopped, and several pointed crosswind a few seconds toward three adult moose 600 m away. Heading inland a few hundred meters single file, they veered for 250 m

until downwind of the moose, then ran straight toward the animals, still browsing when the wolves were within 150 m. Two of the moose sensed the wolves 25 m away and ran. The wolves chased it a few meters until spotting the third moose, which was closer and had not left. They ran the 15 m to this animal and surrounded it.

A few seconds later the moose bolted, and the wolves followed. Soon five or six animals bit at its hind legs, back, and flanks. The moose continued on, dragging the wolves until it fell. In a few seconds it was up but fell again. Arising, the moose ran through open second-growth to a small stand of spruce and aspen, while the wolves continued attacking; one grabbed the quarry by the nose. The moose stood, bleeding from the throat, but the wolves would not attack.

Within a few minutes most wolves were lying down. Two or three continued to harass the moose without actually biting it, and the moose kicked with its hind feet. Whenever the animal faced the wolves, they scattered. It appeared strong and "confident." At 1830 hr we left.

The next morning at 1115 hr the wolves were gone. The moose lay where it had made the stand. It finally arose and moved on. Although walking stiffly and favoring its left front leg, the moose was not bleeding and seemed viable. The wolves were 26 km away feeding on a new kill.

3. March 1, 1959—Isle Royale, Michigan (Mech 1966a)

When we arrived at 1800 hr the wolves had a bull surrounded in a small hardwood stand. He was bleeding steadily from the throat and had difficulty holding his head up. About 15 m² of the surrounding snow was covered with blood. The animal's lower left hind leg was bloody, and he leaned against a tree, keeping his right hind leg centered under him.

Most wolves were meters away, resting and playing but a few licking bloody snow. One, whose legs were covered with blood, harassed the moose, staying near it most of the time, often nipping at the injured leg. However, each time the moose faced it or any nearby wolves, they scrambled away. At 1830 hr we left. Unfavorable weather prevented a check until March 4. At 1045 hr moose bones were scattered where we had last seen the bull. The wolves were 16 km away. This was a bull in wear class VI (8–15 yr old).

4. March 4, 1959—Isle Royale, Michigan (Mech 1966a)

At 1800 hr, 10 of the 15 wolves were traveling along a beach 3.2 km ahead of the others. Several pointed inland toward five moose, the closest 30 m away. As the wolves watched, the moose ran into thick spruces 150 m away, stopped, and looked toward the wolves. Two wolves started toward them a few steps as they disappeared. The area was full of blowdowns and deep snow; the wolves continued on along the shore.

5. March 7, 1959—Isle Royale, Michigan (Mech 1966a)

While five wolves visited a kill on Wright Island, the other 10 traveled into Malone Bay (1815 hr). The leader, 25 m ahead, started toward Malone Island, stopped, and turned toward shore. A few moments later it chased two moose 125 m inland. The moose separated, and the wolf

chased one for 125 m, coming within 30 m. As the moose entered some spruces, the wolf returned to the pack. All wolves then ran to Malone Island.

6. March 7, 1959—Isle Royale, Michigan (Mech 1966a)

At 1820 hr, the above 10 wolves filed onto Malone Island and directly toward a cow and calf lying near the opposite side. The wolves had scented the moose about 400 m downwind. As the pack approached to within 100 m, the cow ran 25 m to the calf. The wolves surrounded the moose but did not attack. Slowly, the moose moved to thicker cover 25 m away. The cow protected the rear of the calf, and several times feinted toward the wolves, which scurried. They lunged at the moose for 4 min but did not attack. Then the wolves headed onto the ice, and lay down. We left at 1830 hr. The next day, tracks showed that the wolves had chased the moose onto the ice, where tracks indicated the moose had stood off the wolves for some time. No blood was seen. The moose had left the island.

7. February 5, 1960—Isle Royale, Michigan (Mech 1966a)

At 1635 hr, we saw the 16 wolves running upwind on an open ridge toward a cow and two calves 1.2 km away. The wolves appeared to have smelled the moose from 2.4 km because they veered sharply toward them at that distance and headed straight for them. When still 1.2 km from them, several wolves pointed toward the moose, which now faced them. The first few wolves charged the moose but a little north of them. Two wolves were far ahead, and two others ran south of the moose's trail. The first two wolves gained rapidly and overtook the moose within 400 m. As the moose ran through open second-growth birch, one wolf remained on each side.

The cow remained immediately behind the calves and twice feinted toward the wolves, which leaped away. Most of the pack began catching up, and as the moose entered a small cedar swamp (the nearest conifer cover); four or five tore at the rump and sides of a calf and clung to it. Within 15 m, the calf dropped. The cow and the other calf continued through the cover with two wolves following for 20 m. When these wolves gave up, the moose returned 50 m toward the wounded calf but gradually drifted toward where they had originally started.

Most wolves concentrated on the wounded calf, which appeared dead within 5 min after falling.

The snow was 30 cm deep, but the wolves sank in about 15 cm.

8. February 7, 1960—Isle Royale, Michigan (Mech 1966a)

At 1700 hr, the pack of 16 suddenly veered upwind and became alert, often stopping and pointing or scenting the wind. All stayed close together and hurried, traveling 1.2 km to within 250 m of a cow and calf browsing directly upwind (1730 hr). The wolves gave no indication of scenting the moose, but when a third of the way through a thick spruce swamp, they suddenly headed toward the moose. When they were within 100 m, the moose fled, cow behind calf. The wolves soon were racing alongside and behind them.

Throughout the chase, the cow defended the calf, charging the wolves frequently. One wolf bit the calf's rump once but to no effect. The pursuit continued 200–300 m through various cover and terrain, and eventually the moose separated. Most wolves pursued the calf, while two followed the cow. After several hundred meters, a few wolves attacked the calf's rump and flanks; one grabbed its left hind leg. The cow caught up and stomped one wolf, which arose instantly and appeared unhurt. The others released the calf and pursued it another 100 m before attacking again. They pulled the animal down and tore at it, but it arose, and the cow rushed in. Some wolves fled, but others chased the cow. Then the wolves assailed the calf again. One grabbed it by the nose, and three or four tore its neck and throat; others ripped its rump. The calf's hindquarters dropped, but the animal dragged its hind legs and the attached wolves. It stood again, and the cow charged again, but one wolf chased her away. The wolves attacked the calf again, and it stayed down. Then they lined up side by side and fed. The cow gradually wandered away.

9. February 11, 1960—Isle Royale, Michigan (Mech 1966a)

The 16 wolves appeared to be following a fresh moose track. When 250 m crosswind of three adult moose (two lying, one standing), they stopped and scented the air (1515 hr). Then they continued along the trail, noses down. Two remained downwind and about 6 m ahead of the trackers. All the moose then were lying, but when the first two tracking wolves got within 6 m, they arose. Meanwhile the rest caught up. One moose ran through a burn, and the others into a dense stand of mature trees; the wolves stood a few minutes.

Meanwhile the single moose, which had run 100 m, started back in an arc toward cover and thus within 50 m of the resting wolves and then reversed itself. The wolves started halfheartedly toward the animal, which continued trotting 800 m and again circled toward cover. The wolves were traveling across the animal's intended trail, apparently having given up. The moose came within 25 m of the two leaders and again ran 800 m away. The wolves tracked the moose 50 m, lay down, and rested for 5 min. The moose circled far behind the wolves toward cover while the pack left.

10. February 12, 1960—Isle Royale, Michigan (Mech 1966a)

As the 16 wolves scented three adult moose 200 m upwind, they started toward them. When the wolves came within 150 m, two moose ran one way and one another. The first two wolves overtook the two moose within 200 m but did not attack. They continued the chase for 800 m through thick, second-growth birch.

The other wolves caught up with the lone moose within 300 m and pursued it another 300, running behind and alongside without attacking. Suddenly the first wolf stopped and lunged at the other wolves, which turned and ran. The single moose continued on. The wolves assembled, and when the two that chased the other two moose arrived, all rested a few minutes.

11. February 12, 1960—Isle Royale, Michigan (Mech 1966a)

At 1340 hr, the 16 wolves headed southwest along a ridge 100 m upwind of a cow (lying) and a calf (standing). The wolves stopped directly upwind and sniffed the wind but seemed unable to determine the moose's location. They stood for several minutes until the cow arose; then they immediately ran to it. The cow hurried to the calf's rear, and the two walked 10 m through an open burn. The wolves followed, but the cow made short charges and kicked at them. Half a minute later the wolves assembled 6 m away while the moose stood and watched. The wolves rested for a few minutes and left.

12. February 12, 1960—Isle Royale, Michigan (Mech 1966a)

The 16 wolves veered inland about 1430 hr toward a lone cow 200 m upwind. It ran when the pack was 100 m away, and the wolves charged and followed her. The cow ran slowly and stopped to look back at the approaching pack, which caught up within 100 m. She stood next to a bushy spruce, and as the wolves lunged, charged and kicked at them with all four feet. Although she seemed to connect with hind hooves, apparently no animal was injured.

Meanwhile, the whole pack caught up. The moose defended herself for 3 min while backed against the spruce, but suddenly she bolted and fled. The wolves attacked her rump and flanks but released her as she brushed through thick spruces. They pursued her for 25 m to where all plunged down a steep slope.

When the moose landed at the bottom, the wolves were attached to her back and flanks, and one held her by the nose. The downed cow tried to rise, but the wolves' weight anchored her. The nose wolf held on firmly while she violently shook her head. Most of the others worked on her rump and flanks, while two tore at her shoulders.

The moose struggled for more than 5 min while the wolves, packed solidly around her, tugged away. Two individuals had to wait at one side, for there was no room. The nose wolf continued its hold for at least 10 min, after which the moose appeared dead. Her tooth-wear class was VI (8–15 yr old).

13. February 15, 1960—Isle Royale, Michigan (Mech 1966a)

At 1410 hr, the 16 wolves suddenly pointed toward a cow 200 m to their left. They then continued down a creek to where it wound closer to the moose. Heading inland over a knoll, the wolves surprised the cow 25 m away.

The moose fled, but the wolves caught right up. One grabbed her right hind leg above the hoof. However, while trotting through some spruces, she shook the wolf loose. She then ran in a semicircle toward a creek, and the wolves overtook her several times but failed to grab her. Once when she ran through a snowdrift, the wolves lost ground, but they quickly caught up again.

As the moose started down a slope, the wolves attacked her rump. She soon shook them and proceeded to a frozen creek bed, where the wolves attacked again. One animal kept jumping at her nose and finally grabbed

it; others fastened onto her rump and flanks. The cow fought hard and dragged the wolves 100 m. Three or four times, she lifted the nose wolf off the ground and swung it several seconds before lowering her head. This wolf maintained its grip for over a minute. The moose finally shook the wolves and ran back upstream with the whole pack following.

The cow started into the woods, and the wolves lunged again. The moose kicked constantly and trampled two wolves into the snow. One crawled away but later seemed unhurt. The moose then stood next to a small balsam and continued to fight off the wolves, which soon gave up temporarily and lay down. At 1435 hr, they assembled 200 m downstream. They returned to the moose three times but found her belligerent, although blood from her wounded rump covered several square meters of snow. Nevertheless, there appeared to be no mortal wound.

From 1450 hr to 1525 hr, the wolves lay nearby. Meanwhile, at 1520 hr the moose walked about 10 m and lay down. At 1525 hr, the pack approached, and she arose again. Although appearing stiff, she charged the wolves effectively. Many were eating the bloody snow where she had stood first. At 1540 hr, the wolves lay back down, and at 1550 hr, so did the moose. A minute later, a wolf approached the moose, and she arose. At 1612 hr, this occurred again. Then the wolves curled up 25 m south of the moose. From 1620 hr to 1640 hr, we were refueling, but when we returned, the wolves were still there.

At 1700 hr, the wolves arose, tested the moose, and found her quite pugnacious. Ten minutes later, 14 of the animals headed southwest, while two remained curled up within 25 m of the wounded moose, which also lay.

From 1735 hr to 1805 hr, the pack visited an old kill 800 m away; the animals then traveled until 800 m from the wounded cow. Meanwhile, the two guard wolves arose and stood near the moose. The pack headed almost directly toward them, and at 1840 hr when we left, it was within 400 m and still heading toward the moose. The next day at 1050 hr the wolves were feeding on the carcass in the same spot.

14. February 22, 1960—Isle Royale, Michigan (Mech 1966a)

At 1505 hr, the 16 wolves were traveling 150 m crosswind of a standing moose. They stopped, milled around, ran

FIGURE 2.2. Wolves tend to attack primarily moose that run from them, although they do not attack all those that run. (Photo by R. O. Peterson.)

back and forth for 5 min, and headed away from the moose for 800 m; then they began backtracking. When they were within 400 m, the moose bolted and ran steadily for >800 m.

At 1550 hr, we left to refuel. The wolves were resting 50 m upwind of where the moose had been. At 1635 hr, the wolves had followed the moose tracks. In one place, they had cut downwind paralleling the trail for 100 m, then veered back to it. They followed the track for 800 m before giving up.

15. February 22, 1960—Isle Royale, Michigan (Mech 1966a)

At 1710 hr, the 16 wolves were 400 m crosswind of a moose browsing in sparse cover. They followed its tracks until one wolf encountered the moose. The moose ran a few meters when the wolf got within 25 m; then the pack surrounded it. The moose stood its ground and charged the wolves repeatedly (fig. 2.1). The wolves left after 5 min.

16. March 1, 1960—Isle Royale, Michigan (Mech 1966a)

At 1615 hr, the 16 wolves cut inland toward two large moose standing 200 m away. As the wolves started up a steep slope, the moose fled for 400 m. The wolves gave up within a minute.

17. March 4, 1960—Isle Royale, Michigan (Mech 1966a)

The large pack (16), at 1200 hr when 150 m downwind of two standing moose, started inland, and the moose ran. One headed around the side of a small lake, but the other avoided it. This moose ran in circles, and the wolves soon overtook it. It stood and threatened the wolves; they left after a minute.

18. March 4, 1960—Isle Royale, Michigan (Mech 1966a)

At 1400 hr, the 16 wolves appeared to be tracking a moose through thick spruces. When within 100 m of two moose, the moose fled. The wolves caught up within 400 m. When the moose split up, the wolves followed the closer and smaller animal; four or five wolves remained close behind and beside the running animal, but the rest were far behind. The wolves caught up and stayed with the moose in the open but soon lost ground in thick cover or blowdown.

Wolves nipped at the animal's heels four times but could not hold on. The moose continued for another 800 m. It stumbled while jumping some downed trees and stopped in a clump of spruces. The wolves (now four) lay near the moose but did not attack. The moose rested a minute, then continued running for ≥1.6 km. The wolves left, and total chase distance was 4 km.

19. March 4, 1960—Isle Royale, Michigan (Mech 1966a)

A few hundred meters farther on, the wolves stopped (1439 hr) and rushed a moose standing 300 m upwind. The animal left immediately, and the wolves gave up without overtaking it.

20. March 4, 1960—Isle Royale, Michigan (Mech 1966a)

At 1610 hr, the large pack (16) followed a fresh moose trail, and the first few wolves approached two large standing moose. The moose stood for 30 sec; then one ran 100 m, stopped momentarily, and continued on. The wolves chased this individual, while the other moose ran elsewhere.

The first moose passed a third lying in an open area, and the wolves surrounded it. This moose stood its ground, and after a few seconds the pack continued following their original quarry, several hundred meters ahead, which was alternately running and standing to look back. The wolves continued 200 m farther (800 m total), gave up, and left. Two wolves at rear almost ran into the third moose but quickly retreated and left.

21. March 4, 1960—Isle Royale, Michigan (Mech 1966a)

The 16 wolves suddenly stopped and headed directly upwind toward a moose standing 75 m inland (1635 hr). It ran hesitantly when the wolves were 50 m away, continued for 25 m, stopped near a small tree, and threatened the wolves. They stood around the moose for 30 sec and left.

22. March 4. 1960—Isle Royale, Michigan (Mech 1966a)

Two moose standing 100 m inland seemed to sense the pack at 1640 hr soon after the previous account. They ran and when the pack was downwind were ≥400 m away. The wolves started toward them but were distracted by three other large moose standing nearby. All three ran, and the wolves chased them all. After 30 sec the wolves focused on one through open cover and gradually heading it toward Lake Superior, which it was paralleling. (One or two wolves usually kept alongside the moose inland.)

After 2.4 km, the moose ascended a small open ridge, and the lead four or five wolves gave up. Those behind took a shortcut and continued the pursuit. This seemed to stimulate the resting leaders, and the whole pack resumed the chase. The moose was then 100 m ahead, and after a few seconds the wolves gave up (1647 hr).

23. March 4, 1960—Isle Royale, Michigan (Mech 1966a)

The 16 wolves, at 1740 hr, veered inland directly upwind toward a cow and calf standing 250 m away. The moose stood, and for a few seconds, both charged the wolves. Then the moose began to run slowly, cow behind calf. The cow continually threatened the wolves, which scrambled away but immediately returned, and the calf charged at least once. After following the cow and calf for 400 m, the wolves gave up at 1845 hr.

24. March 4, 1960—Isle Royale, Michigan (Mech 1966a)

At 1855 hr, the large pack (16) headed upwind toward three standing adult moose. They fled when the wolves were within 100 m. Then one stood, and the wolves chased the other two, which split up. The wolves continued after the closer, a larger animal, staying within a few meters of it. Most had fallen behind, but one overtook it after 1.6 km and stopped it. As soon as the moose stopped, the wolf scrambled away. The moose ran again and all the wolves gave up.

25. March 6, 1960—Isle Royale, Michigan (Mech 1966a)

The 16 wolves left their last kill at 1630 hr and, at 1705 hr, headed directly to a standing moose. The animal stood its ground, the wolves deliberated for 30 sec, and the moose slowly walked off.

26. March 9, 1960—Isle Royale, Michigan (Mech 1966a)

At 1425 hr, the 16 wolves cut inland (crosswind). When 50 m downwind of a cow and calf, they veered toward them. The cow went to the calf's rear, and the two ran when the wolves were 25 m away. Their flight was deliberate and paced, and the wolves followed beside and behind for 3 km.

Whenever the wolves came close to the heels of the cow, she kicked, stopping them momentarily, but they returned immediately. The cow also charged wolves near the calf's rear. Part of the pack stayed beside the animals, and as the cow threatened wolves behind or beside her, others tried for the calf's rump. The cow then charged and scattered them.

Once or twice the calf got 10 m ahead of the cow as she fought the wolves. The wolves attacked the calf two or three times but were driven off by the cow. The calf, which seemed small, chased the wolves ahead of it.

After about 3 km, the moose stopped and so did the wolves (1440 hr). Within a minute the moose were off again, but the wolves remained resting. When 150 m away, the moose began walking, cow ahead.

A few minutes later, the pack halfheartedly started toward the moose again, and the cow returned to the rear of her calf. When within 150 m of the moose, the wolves gave up and rested (1445–1505 hr). The moose continued running for ≥800 m.

27. March 9, 1960—Isle Royale, Michigan (Mech 1966a)

At 1715 hr, the large pack, when 300 m downwind of two large moose (one lying, one standing), suddenly cut toward them. The moose ran when the pack was 15 m away, and the wolves pursued one. It stopped within 50 m, and they continued after the other for 50 m when that moose stopped. Whenever the wolves approached, the moose sent them scurrying. Then it slowly ran a few meters and stopped. The wolves left after 2 min.

28. March 10, 1960—Isle Royale, Michigan (Mech 1966a)

The 16 wolves were heading downwind at 1520 hr, 150 m from a moose lying. They followed a fresh trail nearby leading to the moose. The moose arose when the pack was 50 m upwind and ran 200 m before the wolves found its bed. It continued for 800 m, then stood and watched its backtrail. The wolves tracked for 100 m, lay down, and rested until 1630 hr, then left.

29. March 10, 1960—Isle Royale, Michigan (Mech 1966a)

At 1600 hr, the three wolves started upwind, and when about 50 m downwind of two adult moose lying on the side of a ridge, the wolves ran to the edge and looked over for several minutes. They headed downwind, sat until 1625 hr, turned back upwind, and approached the moose. The closer moose arose when they were 20 m away, and they eventually came to within 10 m. The moose stood for a minute while the wolves watched. Both moose soon ran, but the wolves did not follow.

30. March 11, 1960—Isle Royale, Michigan (Mech 1966a)

The 16 wolves suddenly spread out at 1440 hr and excitedly ran around, 300 m roughly downwind of a cow and small calf. When directly downwind of the moose, they veered toward them.

The moose fled, cow ahead, when the wolves were 150 m away. The pack caught up 150 m from where the moose began, and the calf darted ahead of the cow, where it stayed. Most of the wolves remained in line behind the cow, but a few kept trying to get alongside the animals. The moose passed through various cover types, and the wolves beside the moose never tried to grab it. Twice the moose stopped, but the wolves did not assail them. After about 5 km, the wolves dropped 50 m behind and rested at 1508 hr; the moose continued at least 400 m farther. Each wolf rested where it stopped, but 10 min later they lay together for another 10 min. The snow seemed to hinder the wolves more than usual, perhaps because of a light crust.

31. March 11, 1960—Isle Royale, Michigan (Mech 1966a)

The pack of 16 wolves followed fresh moose tracks at 1601 hr for 1 min, surprising three large moose 100 m away. All three ran, and the wolves pursued one for 50 m until it stopped. Then they chased another, at least 150 m ahead. The wolves tracked it a few meters, and gave up.

32. March 11, 1960—Isle Royale, Michigan (Mech 1966a)

About 1700 hr, the 16 wolves approached two moose lying downwind. The moose sensed the wolves from 100 m and ran before the wolves detected them, but eventually the wolves followed slowly for 50 m. One moose ran directly away, but the other went 150 m, stopped, and watched its backtrail. With wolves 100 m away, it ran another 200. The wolves rested 10 min, and the moose continued on. The wolves then appeared to track the moose but eventually gave up.

33. March 11, 1960—Isle Royale, Michigan (Mech 1966a)

At 1750 hr, the 16 wolves started toward four standing moose 300 m downwind. The moose ran when the pack was within 100 m, and the wolves chased one for 800 m. The moose easily traveled through extensive blowdown that hindered the wolves. Once, the animal charged the wolves, which scattered. It fled and eventually gained 100 m; the wolves gave up (1800 hr).

34. March 11, 1960—Isle Royale, Michigan (Mech 1966a)

After the above episode, the wolves returned to where the chase had begun. A moose stood 150 m away, and the pack arced until downwind of it (1845 hr). Then they approached the moose to within 100 m before it ran. After they pursued for 25 m, the moose stood its ground. The wolves left after a minute.

35. March 12, 1960 Isle Royale, Michigan (Mech 1966a)

At 1115 hr, the pack either scented or trailed two moose standing crosswind. One ran when the wolves came within 125 m, but the other, a young bull, waited until they were 75 m away. Both fled for 400 m, then stood watching their backtrail. The wolves rested where the moose had started, eventually trailed the moose 150 m, but then gave up.

36. March 15, 1960—Isle Royale, Michigan (Mech 1966a)

The large pack (16), when 150 m upwind of two large moose, stopped and pointed. They continued to the moose, which remained bedded until the wolves were within 1–2 m. When the two moose stood, the wolves

surrounded them, but both charged several times, scattering the wolves. From 1513 hr to 1516 hr, the wolves held the moose at bay and then gave up.

37. March 17, 1960—Isle Royale, Michigan (Mech 1966a)

At 1135 hr, the large pack turned upwind and ran about 50 m into thick spruce. Two moose fled the stand and split up. Suddenly a calf, pursued closely by two wolves, left the stand, and within 100 m, the wolves began nipping at its hind legs. After another 50 m, one wolf clung to the animal's rump and the other to its throat. The calf stopped and trampled the front wolf, but the wolf would not let go. It clung to the calf's throat for 2 min while the calf continued to pound it and drag it about.

Finally, this wolf released its throat hold, but the other stuck to the rump. The first wolf then stood on hind legs, and placing its front paws on the left side of the moose, started chewing the side of its neck for several seconds. The calf brushed this animal against a tree, but the wolf then dived under the moose and fastened to its throat. As the running moose straddled the wolf, the wolf ran along with it for a minute.

Meanwhile, two other wolves caught up. One bit the calf's head and finally grasped its nose. The other grabbed its right flank and then clung to its rump for a minute while the moose continued on. Thus, one wolf had the calf by the nose, one by the throat, and two by the rump. The animal soon dropped. In 3 min, it ceased struggling (1145 hr).

When the two wolves first attacked the calf, the cow tried to catch up but was too far behind in the swamp. Total chase distance was 400 m.

38. February 3, 1961—Isle Royale, Michigan (Mech 1966a)

A single wolf, probably a member of the pack of two, at 1700 hr detected two adult moose lying 400 m upwind. The wolf sneaked to within 25 m of one and then ran straight toward it. Both moose fled with the wolf in pursuit. After 500 m, one stopped. The wolf continued after the other but soon fell 15 m behind and gave up.

39. February 3, 1961—Isle Royale, Michigan (Mech 1966a)

At 1800 hr, the wolf in the previous account detected a moose 100 m upwind. It walked slowly to within 15 m, but the moose strode boldly toward it. The wolf "cow-ered," hesitated, circled, and continued on.

40. February 6, 1961—Isle Royale, Michigan (Mech 1966a)

Fifteen of the 16 wolves, at 1630 hr, started to chase three moose, including a cow and calf. The wolves concentrated on the cow and calf, but they had a 150-m start. The lead wolf pursued them for 400 m before giving up.

41. February 6, 1961—Isle Royale, Michigan (Mech 1966a)

At 1750 hr, a few of the 15 wolves sensed a moose browsing 200 m crosswind. The moose detected the wolves 150 m away and fled. The wolves followed hesitantly for 250 m, while the moose traveled 400 m. Then the lead wolf sensed two other moose and abandoned the chase.

42. February 6, 1961—Isle Royale, Michigan (Mech 1966a)

After giving up the previous chase at 1753 hr, the "leader" of the pack of 15 started for two adult moose standing 150 m upwind. Immediately, the moose ran, and the wolf followed. One moose cut to one side and stopped, while the wolf fell 35 m behind the other. The wolf then started for the first. It fled, but the wolf pursued for 150 m. After the moose gained a 25-m lead, the wolf gave up.

43. February 6, 1961—Isle Royale, Michigan (Mech 1966a)

The pack of 15, at 1756 hr, scented a cow and calf browsing 400 m ahead. They continued to within 250 m before the moose sensed them and started off. The moose ran for 400 m, but the wolves did not.

44. Isle Royale, Michigan (Mech 1966a)

At 1145 hr, the 15 wolves scented an adult moose lying 75 m inland, about 400 m upwind. They continued along the shore until opposite the moose. Although the whole pack sensed the moose, only one approached it. When it was 10 m away, the moose arose, and the wolf fled. The action of the pack is unexplained, but possibly the animals had tested this moose 2 d earlier when they last used the route. This was the only moose in the vicinity, and the wolves might have had a recent unsuccessful experience with it.

45. February 24, 1961—Isle Royale, Michigan (Mech 1966a)

Seven members of the large pack sensed, at 1639 hr, two adult moose 75 m upwind. Four wolves remained on a knoll while the other three explored the area trying to locate the moose. When they were within 20 m, the moose bolted, and the wolves floundered through deep snowdrifts in pursuit. The moose quickly gained a 25-m lead, and the wolves gave up (1642 hr).

46. February 24, 1961—Isle Royale, Michigan (Mech 1966a)

The seven wolves just mentioned scented a cow and calf lying 150 m directly upwind at 1752 hr and veered inland through heavy blowdown. Three minutes later, when the wolves were 10 m away, the moose arose. The cow charged, went to the calf's rear, and both walked a few meters. The wolves followed cautiously, but after 30 sec, gave up.

47. February 24, 1961—(Mech 1966a)

A few minutes after the previous account, the seven wolves encountered three adult moose standing. The wolves ran 15 m to the nearest, but this animal stood and threatened the wolves. Immediately, they headed for the second moose, which started running. However, they soon abandoned pursuit, for the animal had a head start. Then they turned to the third moose, which had watched them chase the other. This animal ran on their approach, and when, during the pursuit, it charged the wolves, one got ahead of it. The moose charged this wolf and chased it down the trail for 50 m while the rest of the pack pursued the moose. Finally the moose stood next to a spruce and defied the wolves. Within half a minute they gave up (1804 hr).

48. February 24, 1961—Isle Royale, Michigan (Mech 1966a)

At 1822 hr, the seven wolves scented a moose standing in heavy blowdown 50 m upwind and headed to it. The moose detected the wolves 20 m away, stood its ground, and charged them. They scattered, stood around for 30 sec, and left.

49. February 24, 1961—Isle Royale, Michigan (Mech 1966a)

Three adult moose were standing in heavy blowdown, and at 1835 hr, the seven wolves came within 6 m of the nearest. The other two stood while the wolves chased the closest for 300 m. The nearest the wolves came was 10 m. After the moose gained a 50-m lead, the pack gave up.

50. February 28, 1961—Isle Royale, Michigan (Mech 1966a)

Eight wolves scented three adult moose (one standing, two lying) 35 m upwind (1530 hr). The moose detected the wolves at the same time, and two got a substantial start. The third ran when the wolves were within 10 m. The wolves gave chase but floundered in deep snowdrifts and gave up.

51. February 28, 1961—Isle Royale, Michigan (Mech 1966a)

Seven of the eight wolves scented three adult moose 200 m upwind at 1740 hr. When the wolves were within 50 m, the moose sensed them. Two ran, but the closest stood. The wolves lunged at the moose, and after 2 min, it bolted and the wolves closed in. One grabbed a hind leg, but the moose kicked loose. The wolves chased the animal for 400 m, dropped behind, and gave up at 1746 hr. Deep snow hindered them.

52. February 28, 1961—Isle Royale, Michigan (Mech 1966a)

A moose in the previous account was standing 50 m inland when the wolves approached after the last chase. At 1750 hr, the wolves scented this animal 150 m away. A minute later, when they were 100 m away, the moose strode deliberately toward them for 70 m. When the moose was within 30 m, the wolves left. This is one of the moose that had fled when the wolves first approached.

53. February 28, 1961—Isle Royale, Michigan (Mech 1966a)

At 1756 hr, the seven wolves scented a moose 150 m crosswind, so they cut toward it. They followed its track for 50 m, but when they were 25 m away, the moose ran. The wolves chased it for 10 m, but deep snow hindered them and they gave up.

54. February 28, 1961—Isle Royale, Michigan (Mech 1966a)

The seven wolves scented a moose standing 75 m upwind. The moose detected the wolves about the same time and ran. The pack followed, but only one wolf stayed close. This wolf chased it for 300 m. The moose stopped, and immediately the wolf gave up.

55. March 4, 1961—The Head, Isle Royale, Michigan (Mech 1966a)

The large pack (14 or 15) discovered four moose lying 150 m upwind at 1408 hr. The wolves got to within 50 m before the moose detected them. All the moose ran. The wolves followed for 50 m but did not get close and gave up.

56. March 4, 1961—Isle Royale, Michigan (Mech 1966a)

The 15 wolves, at 1504 hr, discovered the tracks of two large moose standing 150 m downwind. The wolves followed the tracks for 50 m before the moose sensed them. One moose moved toward the other and stood for 30 sec; then both ran. Although the moose had a 30-m start, the wolves came to within 10 m of them at times. After nearly 800 m, the moose curved farther inland, and the wolves abandoned the pursuit. A crust of snow hampered the wolves but not the moose.

57. March 4, 1961—Isle Royale, Michigan (Mech 1966a)

The 15 wolves, at 1621 hr, scented a moose standing 50 m upwind. After they approached it for 10 m, it ran. They followed for a few meters and gave up.

58. March 5, 1961—Isle Royale, Michigan (Mech 1966a)

At 1600 hr, 13 wolves suddenly ran, and 150 m crosswind of them a moose was running. The wolves followed for 25 m and gave up.

59. March 5, 1961—Isle Royale, Michigan (Mech 1966a)

The 13 wolves scented three adult moose standing 200 m upwind at 1744 hr. The moose ran when the wolves were within 150 m, and the wolves chased two, while the third stopped. Part of the pack drove the moose in a semicircle, and the rest intercepted one of them. The other continued running, whereas the cornered moose stood and charged the wolves. They surrounded the animal for a minute and then abandoned it (1746 hr).

60. March 5, 1961—Isle Royale, Michigan (Mech 1966a)

At 1820 hr, the large pack (13) was starting toward a moose when they scented two others lying 200 m upwind. They came within 100 m before the moose detected them. One moose ran, but the closer one stood. The wolves surrounded this moose, but it stood its ground, kicked, and charged, almost connecting with one wolf. Several wolves gathered around the animal's rump, and one grabbed its nose momentarily. The pack spent 5 min harassing the moose. Finally they left (1825 hr).

61. March 5, 1961—Isle Royale, Michigan (Mech 1966a)

The 13 wolves in the previous account scented a moose standing 75 m upwind. When the wolves were within 50 m, the moose ran; the wolves pursued it for about 50 m and gave up.

62. March 12, 1961—Isle Royale, Michigan (Mech 1966a)

A lone wolf was backtracking the large pack to a wounded moose when, at 1223 hr, it sensed a moose standing 35 m upwind. It approached to within 10 m before the moose detected it. The moose walked threateningly toward the wolf, which ran and circled to get by it. Tracks showed that the large pack had, a few days earlier, unsuccessfully tested this animal.

63. March 12, 1961—Isle Royale, Michigan (Mech 1966a)

About 1000 hr, the large pack was at a fresh kill. While backtracking the wolves, we saw a lone wolf also backtracking them at 1030 hr. We continued following the trail to where a badly wounded bull moose lay. Tracks showed that the large pack had wounded this animal, remained for several hours, and left. This probably happened about March 8.

At 1220 hr the lone wolf was 3 km from the moose, still backtracking the pack. From the animal's attitude it appeared to know the wounded moose was there. When 250 m away, the wolf ran excitedly up the trail but became cautious when 50 m away and circled. The moose was lying in a bloody patch of snow 5 m across, which he had not left since the attack. As the wolf came within 30 m, he arose (1255 hr). The wolf approached to within 3 m, circled a few minutes, went off 10 m, and lay down. After 5 min, the moose lay down; immediately the wolf ran to him, and the moose stood. The wolf lay down about 7 m away, and 10 min later the bull lay down. Again the wolf threatened him, tail wagging excitedly, and seemed to try for his nose but failed. The moose just stood without moving quickly or threatening the wolf. A few minutes later the wolf lay down again. The bull continued standing at least from 1320 hr until 1345 hr when we had to leave. From 1445 hr to 1530 hr, the wolf lay on its side about 6 m from the standing moose.

At 1645 hr the bull was alive but lying down and the wolf was tugging at his rump. Intermittently the moose watched the wolf but made no threats. At 1755 hr the moose was still alive, but by 1830 hr he was dead.

64. May 28, 1962—20.8 km Southeast of Palmer, Alaska (Atwell 1964)

At 0432 hr, we observed from a helicopter a 2–3-d-old calf (14 kg) closely guarded by a cow but prostrate. A wolf approached the pair of moose, whereupon the cow, with head lowered, charged the wolf, which then ran in an arc 20–25 m from the calf. The cow halted at the extremity of the arc, and the wolf once more attacked the calf. After 20–30 sec, the cow again rushed the wolf, which, on evading the charge, returned to the calf. At 0439 hr, after eight such attempts (once the wolf picked up the calf by the back and carried it 7 m), the cow gave up, and the wolf fed on the calf. The wolf had pierced its skull, backbone, and left lung.

65. March 8, 1971—Superior National Forest, Minnesota (L. D. Mech)

About 1714 hr, five wolves sensed a moose standing in a clump of conifers and approached it. By 1716 hr, they were within 100 m, and at 1720 hr, one was within 6 m, with that moose looking at it. Two other wolves stood 15 m from the moose at a 90° angle from the first. By 1722 hr, the lead wolf started leaving, and the others followed by about 30 m.

66. May 17, 1971—Superior National Forest (L. D. Mech)

While homing in on Wolf 2258 at 1915 hr, I spotted a moose swimming rapidly south across Cedar Lake. Another wolf arrived around the southwest side of the lake. By then the moose was on the southeast end standing about 6 m from shore in water up to its back. A few minutes later the moose swam north to the middle of the lake. Adult female Wolf 2258 was 200 m west of the lake heading west. The moose swam to the west end of the lake and stood 3 m from shore for several minutes. Then it swam back toward the middle again. We left at 1932 hr. At 2015 hr, when we returned, the moose was back at the west end standing in the lake about 3 m from shore. Wolf 2258 was 3.2 km southwest of there. However, we could not determine whether other wolves were near the moose.

67. March 22, 1972—Superior National Forest, Minnesota (L. D. Mech)

Pack 2415 (seven members), at 1407 hr, suddenly rushed north, each heading directly toward an adult moose. It charged west with both front hooves and stomped on a wolf, first with one hoof, then the other. The wolf ran off. The pack regrouped, and one curled up for a couple of minutes, appearing reluctant to leave. The pack did go on, and after a few minutes, a wolf lay down, but again continued after the pack.

68. February 15, 1987—Denali Park, Alaska (T. Meier)

At 1434 hr, Wolves 235 and 237 were found near an adult moose, which was browsing unconcernedly. The wolves circled behind the moose and grabbed it on a hind leg. The moose did not react much. At 1437 hr the moose looked over its shoulder at the wolves behind it, and they departed.

69. May 22, 1988—Denali Park, Alaska (B. Dale)

At 1339 hr, we found three wolves, which got to within 40 m of a cow and calf moose before they became alert. The wolves walked right by and approached from downhill. They were positioned side by side 3 m apart, and the center wolf was within 10–15 m of the moose. The cow charged, and the wolf fled. After a short stand-off, one wolf lay down or sat while another approached closely from the side. The cow charged quickly, then turned back to her calf. The wolf would follow in close behind her, and the moose would charge again. This happened about 10 times in 7 min before the last wolf gave up.

70. October 10, 1988—Denali Park, Alaska (J. Burch)

At 1050 hr, the Headquarters Pack of seven (Wolves 352, 307) were testing a cow, a calf, and a large bull moose. The cow stood behind the calf and between the calf and wolves; the bull was between the cow and calf and the wolves. The bull appeared to defend the cow-calf pair from the wolves, facing the wolves and charging twice, kicking out at the wolves. The closest the wolves approached was about 2–3 m. The wolves just milled around for 10 min, then left.

71. October 20, 1988—Denali Park, Alaska (J. Burch)

At 1630 hr, Wolf 339 plus seven others curled up on shore, with a moose calf standing in flowing water (air temp around 0°F). No cow seen. We watched for 10 min, but nothing happened. On October 23, C. Roberts and D. Glaser found 339 and seven others leaving kill remains there.

72. June 24, 1990—4.8 km Southeast of Healy, Alaska (T. Meier)

At 2015 hr, Wolf 405 and maybe 343 were near a yearling moose in the center of a 15-m pond. The roiled water implied some activity, and a wolf on shore was shaking. It must have gone into the pond after the moose. The next day, at 1320 hr, the moose was on shore partly eaten, and wolves 405 and 343 were within 500 m.

73. August 6, 1990—8 km North of Healy, Alaska (J. Burch)

At 1003 hr, Wolf 343 slowly approached a cow and calf moose. The cow appeared as though she was aware of the wolf's presence. The wolf slowly closed in. The cow and calf ran (cow in back). The wolf chased them cutting the corners trying to get at the calf, but the cow would always get between the calf and the wolf, and the wolf would stop or change directions. Finally the wolf gave up.

74. October 8, 1991—Denali Park, Alaska (J. Burch)

At 1650 hr, Wolves 5015 and 1080 plus eight others were all resting or sleeping around a standing bull moose. The moose was bleeding from his right hind leg and was standing stiffly. Twice, a single wolf (different each time) leisurely approached the moose to within 15 m, but the moose never moved, and the wolf just stood and stared, then lay down both times. The moose was dead 2 d later.

75. June 22, 1992—Superior National Forest, Minnesota (Nelson and Mech 1993)

At 1000 hr, seven members of the Pike Lake pack were attacking a cow moose running in Arrowhead Creek. The creek was 5 m wide, with curves every 30–50 m. Wolves in the water appeared to be swimming. Some ran along each creek bank as the moose ran and swam down the center. Others jumped toward the moose from each bank and grabbed at what they could. At least one time, all seven wolves were hanging on the moose.

One wolf held the moose's nose and was thrown from side to side by the thrashing moose. Others grabbed the shoulders, flanks, and rump but were shaken off within seconds by the moose's momentum and running. One wolf was atop the moose crossing a deep section of the creek where the moose appeared almost fully submerged. Individual wolves swam to the nearest bank, crawled out, and ran to the next creek curve where they crouched and waited for the moose.

Once the moose was opposite them, the wolves leaped into the air, landing just short of the moose. Once, the moose momentarily flailed at a mostly submerged wolf with its front legs. After traveling about 500 m, the moose abruptly reversed direction, but the wolves continued the attack. At 1024 hr, the moose stood in shallows as the wolves ceased and rested on one bank. The moose moved into deeper water but immediately returned to the shallows. It appeared steady and strong with no large wounds or blood visible. We left at 1042 hr. Four hours later the moose was gone, but the wolves remained bedded at the site. Three days later, this pack was 10 km distant, and there was no evidence of a dead moose.

76. August 11, 1992—Denali Park, Alaska (A. LaFever)

I had several reports from visitors in Teklanika and Igloo campgrounds (0830–1030 hr) about four wolves north of Teklanika Bridge stalking a cow moose. The wolves chased the moose, and then she charged them. They ran into the trees, and the cow stood in the river, knee deep. One wolf was seen biting the moose's left rear leg, and much blood spurted out. The moose kicked a wolf, and it seemed in much pain and was not using one leg. I saw the cow about 1130 hr, but no wolves. The moose's rear left leg was bloody. The moose continued heading north. Outcome unknown.

77. June 3, 1993—Copper River Delta, Prince William Sound, Alaska (Stephenson and Van Ballenberghe 1995a)

We aerially viewed three wolves attacking a cow moose with neonatal twins (approximately 1 wk old). Of the three wolves, one was a radio-collared, black female wolf (Wolf 1), one was a radio-collared, gray female (Wolf 2), and one was a noncollared gray (Wolf 3; sex unknown). In March 1993, the gray female had a preexisting severely dislocated tibiotarsal joint, with bone protruding through skin. Based on her age and future locations at a den (with pups) she appeared to be the breeding female. In March, the black female was either a pup or yearling.

At 2021 hr, the three wolves circled a cow moose with twins and alternately and/or jointly rushed them. The cow diligently kept the twins together and defended them by rushing the wolves and kicking with her forelegs.

After 2 min, one wolf rushed in, grabbed one calf, and knocked it down but was immediately chased away by the cow. The wolves continued and were periodically pursued by the cow less than 10–20 m from the twins, and the cow may have occasionally lost sight of the twins. However, primarily she stood over the twins. As time passed, the frequency of the cow's pursuit of the wolves increased as did the cow's separation from the twins. Occasionally one calf tried to follow the cow.

At 2031 hr, while the cow chased a wolf, one calf fled toward a stream. The wolves appeared not to detect this movement, but Wolf 2 began to follow it about when the calf reached the stream. As the calf swam across, the wolf quickly swam after it. The wolf soon reached the calf, but it swam downstream about 30 m before the wolf reached it again. Then, the calf escaped and swam upstream 10 m before the wolf seized its neck and pulled the calf ashore.

The wolf then began consuming the calf, which was still moving its legs. During the wolf's pursuit of the calf, the two other wolves continued to harass the other moose. The cow, now with one calf, kept the calf under her, protecting it. Wolf 2 dragged the dead calf about 50 m and continued feeding.

At 2045 hr, Wolves 1 and 3 ceased intensive harassment of the moose but remained within 40 m except that Wolf 1 headed to where Wolf 2 was feeding. Wolf 1 then returned to the cow and calf moose. Meanwhile, Wolf 3 sat and observed the moose from 50 m. As Wolf 1 returned, the cow forcefully repelled her.

At 2054 hr, the cow began to move her calf toward the stream, but after 30 m, Wolf 3 resumed pursuit. The moose retreated to taller cover. Shortly thereafter, all three wolves harassed the moose. However, after another 2 min, the wolves left the immediate vicinity.

At 2108 hr, the cow began to depart for the stream, but the calf appeared reluctant. However, after 2 min, the cow and calf walked to the edge of the water. Two minutes later, Wolf 2 walked around the moose, which remained in the stream. Continuous observation ceased at 2117 hr, but the moose remained in the stream.

Wolves 1 and 2 remained near the cow and calf up to 2230 hr, and Wolf 1 was mildly harassing the cow from the bank. The cow rushed from the water, chased Wolf 1 away, and returned to the calf. Wolf 2 was not visible. The cow and calf were still standing in the stream when we departed at 2235 hr.

78. August 24, 2008—800 m West of Saginaw Point, Isle Royale, Michigan (Jordan et al. 2010)

We were on a bluff overlooking an unnamed, 1.7-ha pond about 26 m below and heard the moaning of a lone, adult cow moose swimming toward the west shore. She then stood at the edge while foraging on aquatic plants in a depth of about 1.75 m. Her hindquarters were about 1 m from the bank. Another lone cow, probably a yearling, stood above on flats 40 m farther from us.

Three adult wolves ran toward the cow and leapt into the water atop the moose. The cow headed away from shore, apparently swimming, although her hooves were stirring up substrate. The wolves were swimming except that at least two repeatedly mounted the cow's back, each time for a few seconds, and seemed to be biting at its neck. The frequency of these mountings increased during our observation. When a wolf reached the base of the neck, the cow expelled it by vigorous side-shaking of its head and neck. At first, wolves were swimming less than 5 m behind the moose but steadily closed to 1–2 m.

From the middle of the pond, the cow first headed to within 1 m of a point where going ashore would be readily feasible, but then turned back toward the middle. At that point, one wolf that had been trailing the others took a few steps onto shore but jumped back in when the cow turned back. From there the attack progressed to the center of the pond and out of view.

(During the chase, the yearling followed on land, apparently watching the chase. About 5–10 min later it returned to near its initial location.)

One week later on shore toward where the chase had been heading we found a well-eaten carcass of a wolf-killed cow, 14 yr old, with significant arthritic degeneration in the pelvic region.

Conclusion

It should be clear from the above observations that moose defenses are many, varied, and highly effective against wolves. Thus wolves have all they can do to overcome those defenses, and they risk their lives in the process. Where moose make up the only important food of the wolf, such as on Isle Royale, Michigan, in Lake Superior, wolves spend just about all their waking hours—when not actually feeding—searching for, chasing, or confronting, catching, and killing moose. For example, during 31 d from February 4 to March 7, 1960, a pack of 16 wolves covered 443 km and killed four adults and seven calves (Mech 1966a). At a mean travel rate of 8 km/hr (Mech 1966a, 1994b), they averaged spending 5 hr and 40 km hunting per moose killed.

We now know that the primary piece in the puzzle of how wolves are able to overcome such effective defenses is moose condition. Moose defenses work fine for most of the animal's life, probably hundreds of times per moose. However, when moose pass their prime of life or become debilitated in some other way that is when wolves can finally break through their defenses and kill one. Thus wolves must make many attempts to kill moose for each success. Based on 115 direct winter encounters between wolves and moose on Isle Royale, wolves succeeded only 6% of the time (Mech 1966a; Peterson 1977). In Denali National Park, Alaska, from 1969–70 to 1973–74, wolves succeeded in killing only 6% of the 346 moose they tested during winter (Haber 1977). The comparable figure for Denali Park in 1986–93 was 13%–26% success, based on 53 moose tested and depending on whether moose that were only wounded during an observation later died (Mech et al. 1998).

The reported success rate of wolves hunting moose in Scandinavia was 45%–64% (Sand et al. 2006). The explanation given for these figures being so much higher than those for other areas was that the Scandinavian moose studied were naive to wolves, which had only become re-

FIGURE 2.3. Usually wolves tend to bite the rump of a moose first and most tenaciously. (Photo by R. O. Peterson.)

established 10–15 yr earlier (Wikenros et al. 2009). However, the Isle Royale wolves had also only inhabited that island for about 10 yr before the Isle Royale data were collected. Other explanations for the different rates may be that the Scandinavian data were gathered by snow tracking on the ground, where it is easier to find successes than misses (Kolenosky 1972; Carbyn et al. 1993), and that a high percentage of their kills were calves.

Evidently only strong moose in their prime and in good condition stand their ground, while old or debilitated moose tend to run; when moose do run, wolves can somehow pick out those that show some weakness. Wolves are very good at detecting subtle visual cues (Range and Viranyi 2011), and the fact that dogs can detect cancer by odor (Willis et al. 2004) makes it seem reasonable that wolves might be able to smell prey diseases or disabilities as well. For example, moose sometimes develop necrotic mandibles/decaying teeth associated

with actinomycosis (Mech 1966a), and R. O. Peterson (pers. comm.) suggested that wolves might be able to smell that condition.

To find such an old or debilitated moose, however requires wolves to travel far and wide and check out many moose, as the wolf's low success rate indicates. For example, an Alaskan pack sometimes covered 72 km/d (Burkholder 1959). On Isle Royale, a pack of 16 wolves averaged traveling 50 km/d from February 4–March 7, 1960, when they were actually hunting (excluding time around kills [Mech 1966a]). A pack of at least 22 wolves in northern Canada sometimes traveled a minimum of 41 km/d while hunting musk oxen (Mech and Cluff 2011).

Some moose escape unscathed while others are wounded but escape, at least some to be killed later. Still others are slowed by the attack. The latter continue to fight, but wolves bite at the upper hind legs and try to attach to the rump of the moose (fig. 2.3) or tear pieces out of it (Mech 1966a; Buskirk and Gipson 1978). Often one wolf manages to latch its fangs into the rubbery nose of the moose and hang on. Mech has seen moose swing such a wolf around in the air, but the wolf maintains a solid grip (account 13). In other cases, wolves actually ride on the back of moose (accounts 75 and 78). Meanwhile, other pack members continue to tear out the rump of the moose and possibly the large intestines until the moose succumbs, which often takes hours or days (see Mech 1966a, fig. 96). Often (seven out of 14 cases in Denali, 1986–93), wolves only wound the moose and then give up. Presumably they return and eventually kill the wounded moose as the weakened animals become increasingly vulnerable.

On Isle Royale, a single wolf killed a moose wounded by a large pack an estimated 4 d before (account 63; Mech 1966a, figs. 94, 95). Single wolves can even kill a moose that is not wounded, although it may take 2–3 d (Mech et al. 1998, fig.5.8). On Isle Royale in winter five radio-collared lone wolves killed moose, and one of the wolves killed more than five moose (Thurber and Peterson 1993).

Again, the key to wolf success, whether it be a single wolf or a large pack, is the vulnerability of the moose. Although moose or other prey can be vulnerable because of a wide variety of conditions, probably the most common, documented vulnerability of adult moose is being young (especially calves) or old. The most extensive data about ages of adult moose killed by wolves come from Isle Royale where only 36% of 307 adult moose killed by wolves were less than 6 yr old (Mech 1966a; Peterson 1977). However, similar findings have been made in several others areas, with beginning age of vulnerability varying by population (Haber 1977; Fuller and Keith 1980; Peterson et al. 1984; Ballard et al. 1987; Bjorge and Gunson 1989; Hayes et al. 1991; Gasaway et al. 1992; Mech et al. 1998; and Mech and Nelson 2013). In the most detailed analysis of the relationship between age of moose vulnerability to wolf predation and its effect on the wolf population, Peterson et al. (1998) and Vucetich and Peterson (2004) found that the abundance of moose at least 10 yr old explained 80%–85% of the variation in wolf numbers on Isle Royale.

As for factors predisposing moose calves to wolf predation, that question is only now being investigated (G. D. DelGiudice and W. Severud, pers. comm.). Possible predisposing factors based on studies of other ungulate young include low birth weight, deviant blood values indicating physiological/nutritional problems, and maternal condition and experience (Barber-Meyer and Mech 2008).

In any prey population, individual variation will almost always leave some individuals more vulnerable than others. The wolves' constant scanning and testing of these populations allows the carnivores to find those individuals and cash in on them.

3

Caribou

CARIBOU, AND THEIR Old World equivalent, reindeer, inhabit the arctic and subarctic of Canada, Alaska, Greenland, Fennoscandia, and Russia, from about latitude 55° to 85° north. Barren-ground caribou inhabit the tundra and barren grounds; woodland caribou inhabit the taiga; and mountain caribou, mountains. A smaller version, the Peary caribou occupies the High Arctic of Canada, but all caribou are of the same species (Hummel and Ray 2008). Both sexes sport antlers, although female antlers are much smaller than those of males. Females weigh 55–110 kg and males 90–180 kg. Newborn calves weigh 7–9 kg. Caribou graze on lichens, sedges, forbs, and low woody plants, stripping leaves of branches while always on the move. Barren-ground caribou are herding animals that generally live nomadically and, in North America, are seasonally migratory. Wherever caribou live, wolves live nearby; when caribou are migrating, wolves are not far behind. All the available observations of wolves hunting caribou are based on barren-ground caribou not only because they are far more numerous than woodland caribou but also because they live in the open, where biologists can more easily observe them. Thus, this chapter deals primarily with barren-ground caribou.

Woodland caribou live more like white-tailed deer; presumably their interactions with wolves are similar to those of whitetails. Like whitetails, woodland caribou tend to "space out" (Bergerud et al. 1984) to calve and to "space away" from wolves in general (Bergerud 1985). Even in winter, caribou locations averaged 15 km from wolves in Ontario (Schaefer 2008). Also like deer, woodland caribou group into herds during winter in areas of less snow but their herds only reach up to a few dozen (Schaefer 2008), whereas whitetail herds often number many more.

Much is known about wolves hunting caribou not only because of Murie's (1944) early studies in Mount McKinley (now Denali) National Park but also because of a spate of barren-ground caribou studies conducted in Canada and Alaska in the 1950s and 1960s. Caribou were economically important for both subsistence and sport hunting, so Banfield (1954) and Kelsall (1957) investigated caribou herds in Canada's Northwest Territories. Because caribou form the main prey of wolves in that area, caribou studies automatically involved the interactions between caribou and wolves. About the same time as the Canadian caribou studies, Crisler (1956) published important observations of wolf-caribou interactions in Alaska. Kelsall (1968), Mech (1970), and Mech et al. (1998) synthesized much of the information from those studies, and Bergerud and his associates contributed considerably to the later caribou literature, especially that of caribou antipredator behavior (Bergerud 1985; Bergerud and Page 1987; Bergerud et al. 1984, 1990). Although caribou, like all of the wolf's prey, possess several antipredator behaviors and strategies, probably the most effective is the creature's tendency to herd. Caribou evolution has honed this strategy to its highest form, with caribou herds often numbering in the thousands. As already described with white-tailed deer (chap. 1), herding affords many protective advantages. However, not all caribou inhabit herds all the time. For various reasons, some individuals sometime "straggle" (live apart from) herds. At such times that behavior might even be adaptive, for wolves might spend most of their time with a herd once they have found it.

Healthy caribou have several backup antipredator strategies as well. The most important is nomadism. Caribou tend to roam nomadically throughout much of the year, constantly moving as they graze. During spring and fall many herds migrate between tundra and taiga. While migrating, they travel at about 7.2 km/hr (Pruitt 1960) and can cover about 20–65 km/d (Banfield 1954). Such constant and long-range travel makes their location much less predictable to wolves trying to find them.

During calving season, caribou movements are especially protective of newborn calves. Such calves are highly vulnerable during the first few weeks of life, and commonly half the calves perish before they are 6 mo old (Bergerud 1980). In Denali Park from 1984 to 1991, wolves killed 12%–34% of the calves less than 16 d of age each year (Mech et al. 1998). It is not surprising, then, that caribou have evolved various strategies that help minimize such losses. One such strategy involves the synchrony of calf births (Adams and Dale 1998). With calves being so vulnerable for their first few weeks, wolves and other predators can easily make a quick killing on them. However, while the carnivores are busy doing so, time is elapsing, and surviving calves are developing and growing stronger. If the vulnerability window is short, a phenomenon known as "swamping," more calves can reach an age where they become much harder to catch. If the calving period were prolonged, wolves would have a longer period to find the most vulnerable calves and thus take a higher percentage of them.

In some areas, caribou cows move to islands and shorelines to calve (Bergerud 1985). In both cases, the caribou not only are farther off the literal beaten path but also near water, which provides a measure of safety (see accounts below). In other areas, caribou move to high mountain tops or extensive spruce swamps isolated from other prey and predators (Adams et al. 1995). In Denali Park, caribou calved in an area more or less outside of the most heavily used parts of two wolf-pack territories (Mech et al. 1998). During most of the year that area was devoid of prey, so wolves used it little. One of those territories was unstable over a 4-yr period, with at least three of its resident wolves starving to death. Even if wolf use of the calving ground was less for only a few critical weeks, that difference could significantly improve the calf crop's chances of survival. Caribou in northern Canada use this spacing-away strategy to the extreme by migrating hundreds of kilometers away from other prey where, because of lack of prey most of the year, wolves are scarce (Bergerud and Page 1987). By the

time wolves have found the bonanza of calves, most are less vulnerable.

As calves become more mobile and less vulnerable, their cows join others and gradually form nursery herds of increasing numbers of cows and calves. At this time the calves begin enjoying all the advantages of herding enumerated above, and their vulnerability per calf drops considerably. A single vulnerable calf in a group of 100 possible targets has a far greater chance of survival than if alone. By early July, nursery herds can number 1,000–10,000 (Kelsall 1968).

An important backup defense that caribou excel at is speed (Mech 1970; Haber 1977). Caribou sensory abilities are not very sharp (Kelsall 1968), so running ability is even more important to caribou than to white-tailed deer. Capable of running at 60–80 km/hr (Nowak 1991), caribou can often outrun wolves, whose speed has been estimated at up to 64 km per hr (Mech 1970). Even calves, after a month or so of age, can keep up with their mothers for certain distances. Calves grow and develop quickly, quadrupling their weight by mid-August (Kelsall 1968).

Caribou tend to calve in mid-May to mid-June in Denali, and Murie thought 7–10-d-old calves could run almost as fast as their mothers. On June 4, Murie observed a caribou herd being chased by a wolf, and "the cows and calves fled full speed along the base of the mountain, a calf leading the flight." On June 29, Murie (1944) watched a male wolf chase a herd of 50 cows and calves (4–6 wk old) for about 800 m before a calf started falling behind (account 9).

The calves in both cases might not have been able to run at top speed yet, for presumably the cows would only have been running as fast as their calves could go, not necessarily at their own top speed. As it was, the wolf was closing in on the herd in the latter case (even after once tumbling head over heels!), an indication that the herd was not fleeing as fast as it could have without calves. Nevertheless, it is clear that fairly young calves were quite capable runners already.

A 2-mo-old calf far from its mother was observed running so fast from a wolf that it overtook its fleeing mother (account 16). Of course, calves do not reach adult size until about 2–3 yr (Kelsall 1968), so their shorter legs and poorer reserves will prevent them from reaching adult speed and endurance, thus making them more vulnerable than adults for quite a period. Nevertheless, in Denali Park, wolf capture of calves dropped off consistently after about 3 wk after calving began (Adams et al. 1995).

As adults, caribou can run so fast and for so long that most of the time they escape wolves. Murie (1944, 164) observed that, "in a short chase, the grown animal can run away from wolves." Even when surprised only 30 m from a wolf, a caribou can use its speed to save itself (account 17). Because healthy caribou can rely so much on their speed, they often allow wolves to come fairly close. Banfield indicated that "caribou farther than 300 m from the wolves ignored them," and healthy adult caribou sometimes let wolves get to within 75–100 m before fleeing (account 13). In such cases, the wolves may make a cursory attempt at their potential prey but also give up quickly. Similar to deer, caribou stop running soon after they realize wolves have abandoned the chase, and no doubt for the same reason—to save their strength for when they are hard pressed and really need it. In some cases, however, caribou flee just because they see others in flight. This behavior might result from the caribou's poor senses. If they aren't sure where their adversary is but see other caribou fleeing, they realize danger is near, and until they can size up the situation, the safest strategy is to leave.

It appears that caribou can distinguish when wolves are targeting them versus when they are just passing through the area based on the wolf's profile. When merely traveling, a wolf holds its head up such that in a head-on profile, the head shows above the back; a stalking wolf lowers its head and points it forward such that the wolf's back is just above the head in direct profile (Pruitt 1965).

Although on a straight-out run, caribou can usually outdistance wolves quickly, conditions and circumstances sometimes thwart caribou's speed because most terrain, except frozen lakes, is not flat and level. Deep snow can also thwart caribou escape, especially drifted snow in tundra depressions. Thick, brushy vegetation can do the same, although such conditions can also hinder wolves as well or even more, again depending on the exact circumstances.

When such circumstances allow wolves to catch up to caribou, sometimes the quarry can reach water. Lakes, ponds, rivers, or streams often afford caribou (and other wolf prey) adequate protection (account 16). With longer legs, caribou have the advantage in both running

through water and standing in it. Even shallow water can provide an advantage to caribou. "The [river] crossings didn't slow the caribou, but drastically slowed the wolf" (account 48).

When all the above defenses fail or when circumstances favor wolves, caribou possess a final defense: their antlers (Miller 1975). In Denali Park, observer B. Looney reported the following: "The caribou stopped in the middle of the river, turned around and faced the wolf. The wolf came up to the caribou and made side to side attacks at the caribou, darting toward the right and left side of the head. At each attack the caribou lowered its head, placing the 'shovel' part of its antlers in direct path of the attack" (account 48).

Although we know of no documented death of a wolf from caribou antlers, at the very least the antlers can act like a foil as in the case above. Furthermore, if the sharp tines of a white-tailed deer can kill a wolf (Nelson and Mech 1985), it seems possible that so could caribou antlers. Possibly caribou hooves, too, could disable or kill a wolf, although those hoofs are duller and blockier than those of deer, moose, or elk. Caribou do use their hooves for self defense, however. At least some caribou cows tend to rear up and "box" humans trying to catch their calves and try to trample anyone who persists and happens to get in their way (Mech et al. 1998). Presumably they would do the same to wolves.

The bottom line is that, except for calves less than 3 wk old, most caribou survive wolf encounters most of the time, as is the case with all species of wolf prey. Estimated success rates for wolves hunting caribou in Denali Park vary from 1% to 10% of total caribou chased (including those in herds). Even when attacking groups of caribou, wolves succeeded in making a kill only 15%–56% of the time (table 3.1).

Hunting Accounts

The following accounts of wolves hunting caribou are paraphrased from various publications but also include several unpublished descriptions from edited field notes. These accounts include almost all of those available, although space constraints require that we not include those with the least amount of new information. A particularly informative depiction of wolves hunting caribou can be found on YouTube at http://www.youtube.com

Table 3.1. Success rates of wolves hunting caribou in Denali Park, Alaska

	No. of Caribou			% Success Based On	
	Individuals[1]	Groups	Killed	Individuals	Groups
Summer[2]	1,934	110	54	6	49
Winter[2]	86	16	9	10	56
All year[3]	303	26	4[4]	1	15

[1]Includes cow and calves.
[2]Synthesized from Haber 1977.
[3]Mech et al. 1998.
[4]Three were newborn calves.

/watch?v=AoE6geAq1k8 ("Wolves Hunting Caribou," from "Great Plains," episode 7 of *Planet Earth* [BBC, 2006]). All distances are estimated. Many observations were made in Denali National Park and Preserve (former Mount McKinley National Park), and the locations have been abbreviated to "Denali Park." Numbered wolves are part of a radio-tracking study (Mech et al. 1998), and their ages and sex are given when no nonradioed wolves accompanied them.

1. June 16, 1939—Denali Park, Alaska (Murie 1944)

Around noon we noticed about 250 caribou galloping, mainly cows and calves. A black wolf was galloping after them. When the caribou reached flats in front of us, the wolf was close on their heels. The rear caribou fanned out so they were deployed on three sides of the wolf. It continued straight ahead, continuously causing those nearest to fan out, making an open lane through the herd. Those on the sides stopped and watched the wolf go past. Soon most caribou were on either side of the wolf's course.

The wolf stopped for a moment, and so did the caribou. Then the wolf continued straight ahead after a band of about 30, and these again fanned out, whereupon it swerved to its left after 15 of them, which then started back toward where they had come from. The wolf chased them about 50 m and stopped. Small bands, some only 100 m away, almost surrounded it. Then the wolf started after 25 cows and calves farther away. Before they got underway it gained rapidly. For a time the race seemed quite even, and I felt sure the band would outdistance their enemy. I was mistaken.

The gap commenced to close, at first almost imperceptibly. The wolf was stretched out, long and sinewy, doing its best. Then a calf dropped behind the fleeing band. It could not keep the pace. The space between the band and the calf increased while that between the calf and the wolf decreased. The calf lost ground more rapidly. The wolf seemed to increase its speed and rapidly gained. When about 10 m ahead of the wolf, the calf veered from side to side to dodge it. Quickly the wolf closed in and at contact the calf dropped about 50 m behind the herd. The wolf seized it at the shoulder. They had covered about 500 m.

2. June 17, 1939—Denali Park Alaska (Murie 1944)

A band of 15 adult caribou trotted along followed by a trotting gray wolf. When the wolf stopped, the caribou stood watching from 75 m. The caribou left, and the wolf disappeared.

About 1000 hr, eight or nine cow and four calf caribou galloped across a river, followed by a loping gray wolf. The fleeing caribou, with a long lead, reached rough country just below a ridge. The wolf gained while the caribou ran up and down the slopes and it ran on the level, but when the wolf also reached the rough country it quickly fell behind and stopped after >800 m.

3. June 1, 1940—Denali Park, Alaska (Murie 1944)

A black male wolf moved down a slope at an angle toward some caribou and got to within 250 m before they fled. He galloped hard up a low ridge and into a shallow ravine where he captured a calf after following it in a small half circle.

4. June 1, 1940—Denali Park, Alaska (Murie 1944)

All day caribou had been in the vicinity of the den, but five resting wolves nearby did not molest them. Soon after three male and one female wolves departed, caribou ran off in various directions having seen them.

Small bands of caribou were scattered over the tundra. The black male wolf was far ahead of his companions. As usual he seemed to be doing most of the hunting. He approached two or three bands and watched while they fled. There happened to be no calves, so I wondered if the wolf was surveying each band for calves. The two grays caught up with the black male. Once, the large, black-mantled male dashed at a band, then stopped to watch. The scattered caribou assembled and ran off. No calves. Then the black male galloped after a herd but stopped to watch when he was near it. After traveling 8 km, the wolves were together again and had made no serious effort to kill caribou. The bands of 15 or 20 caribou the wolves had encountered had no calves, which the wolves seemed to be searching for. Some bands that ran from the wolves veered only a few hundred meters. Others that went straight ahead of the wolves ran as much as 1.6 km. Some fled because they saw others run, and a few came nearer to the wolves.

5. June 17, 1940—Denali Park, Alaska (Murie 1944)

At 1730 hr, the black female wolf appeared south of me, about 1.6 km from the den, chasing a large band of caribou containing many calves. Some caribou veered off and began to feed. A calf brought up the rear of a group the wolf was chasing. When it appeared the wolf might overtake the calf, most of the band and the calf veered upward to the left and seemed to increase their speed. The wolf singled out another calf running straight ahead with four or five adults, but in a moment the chase went out of sight.

6. June 19, 1940—Denali Park, Alaska (Murie 1944)

About 800 m north of the den, at 1100 hr, one of the black wolves chased 35 cows and calves for 400 m, then gave up.

7. June 22, 1940—Denali Park, Alaska (Murie 1944)

At 2035 hr the black-mantled gray male and the black male followed the river bar south from the den. About 5 km away, 200 or 300 caribou fed on a flat. For about 1–3 km the wolves trotted together; then the gray fell far behind. He moved along the east bank while the black trotted briskly toward the caribou. When 200 m from them, he watched for a minute, then started galloping. He drove all the caribou off the flat toward a gravel bar.

He did not try to catch them but was definitely herding the scattered animals. When he had run the length of the scattered herd and had the caribou all galloping out he swung around in front of the herd and came back chasing them. As he caught up with a band it would veer. Then he continued straight ahead to the next little band which in turn would veer. Finally he wandered around as though investigating the area, then trotted off. The caribou moved on as though they had completely forgotten the chase. The black-mantled male wolf had lingered across the bar. Some caribou that had been driven out on the bar had drifted near him, and he had chased a band.

Had the hunt followed a general pattern of cooperative maneuvering? If the black had chased the caribou toward the gray far enough to tire the caribou, then the gray wolf could have taken up the chase fresh.

8. June 23, 1940—Denali Park, Alaska (Murie 1944)

At 0930 hr, 250 or more cows and calves were running hard, with a wolf chasing them. The wolf chased one group after another so that he had the various groups running in different directions. Although galloping hard, he did not bear down on any herd. It looked like he was testing the groups for an especially vulnerable calf. After considerable chasing, the wolf ran after four adults and a calf, driving them off by themselves. The calf broke to one side and kept veering as though trying to return to the herd but in so doing lost ground, for the wolf then cut corners. When the wolf was about 20 m behind the calf, it was unable to reduce the gap for some time, but when the calf zigzagged it lost ground. The wolf gradually closed to a few meters, but still the chase continued for another 200 m. I thought the calf might escape, so well was it holding up. But the wolf finally closed in and downed the calf. While the wolf was apparently biting it, it jumped up and ran 75 m before again being overtaken. After disposing of the calf, the wolf trotted toward the herd, then returned to his prey.

9. June 29, 1941—Denali Park, Alaska (Murie 1944)

At 1500 hr, a band of 400 caribou were running over the tundra 1.6 km west of the den, and the black male wolf first ran toward one end of the band. The herd broke into groups of 50 or 60, and the wolf dashed along in the middle. Then the wolf started after 50 cows and calves for 800 m, with the wolf closing in. Once he stumbled and rolled completely over. But he was up quickly with little time lost. Then a calf dropped behind the others. This seemed to encourage the wolf to run faster, and in <400 m he overtook the calf, knocking it over more readily than usual.

10. June 29, 1941—Stony Creek, Denali Park, Alaska (Murie 1944)

Foreman Brown of the Alaska Road Commission camp watched a wolf with a crippled hind leg chasing calves. After chasing some bands without success, the wolf moved off and waited for the herds to approach. But while the caribou were some distance away it gave chase. It was fast for a short distance but quickly tired and fell behind. It caught no calves while Brown was watching.

11. July 19, 1941—Denali Park, Alaska (Murie 1944)

At 1000 hr, the black female wolf was on the east Fork River bar circling back and forth with her nose to the gravel. She made a sweep of 100 m downstream, returned upstream, and waded the river. After crossing, her ears were cocked forward, and she started on an easy lope, apparently focused on a definite point. Ahead of her lay a caribou calf watching the wolf. When the wolf was 150 m away, the calf jumped up and galloped upstream, crossing and recrossing the creek a dozen times, and every time the wolf followed. The calf seemed slower than younger ones, for the wolf was running easily and gaining. Some 800 m from the start, the calf, now hard pressed, wallowed in a deep part of the stream, stumbled, and fell. In a jump or two the wolf caught up and pulled the struggling calf, which once gained its feet for a moment, across the deep part of the stream to the shore, where she quickly killed it. This calf was a straggler and therefore may have been a weakling.

12. August 4, 1948—Clinton-Colden Lake, Northwest Territories, Canada (Banfield 1954)

At 0830 hr, a large group of caribou became agitated, many peering toward a knoll, and others trotting about restlessly. A large, white wolf came out from behind a knoll. The caribou fled in a compact group of about 50. Both wolf and caribou maintained top speed for about 100 m, and the wolf seemed to be losing ground. It quickly gave up. Other bands to the flank ignored the chase.

13. August 11, 1948—Angikuni Lake, Keewatin District, Northwest Territories, Canada (Banfield 1954)

Lawrie and Mowat observed the hunting tactics of three wolves in line abreast 200–300 m apart in view of eight scattered bulls, which they passed at 200–600 m. Caribou >300 m away ignored them, even though the wind was favorable. The nearer caribou snorted, jumped a few paces, and watched the wolves passing. The wolves later drew together, then spread out again. While its companions watched, one flushed a calf from a willow clump 100 m away. The wolf gave pursuit. In the first 50 m, a cow and calf flushed, and these ran beside the wolf on a converging course until about 15 m separated them. The wolf then gave up on the first calf and swung toward the second, stopping after only 100 m. The first calf had stopped and watched. A second wolf then trotted toward it. When the wolf was 100 m away, the calf fled. The second wolf sped up for 25 m, slackened to a walk, and sauntered on. Meanwhile the cow and calf circled back and passed within 100 m of their former pursuer, which ignored them. The wolves continued abreast across the swale. One stalked a yearling, which let it approach within 75 m before fleeing in a circle. The wolf did not follow but joined its companions.

14. August 14, 1948—Clinton-Colden Lake, Northwest Territories, Canada (Banfield 1954)

Wilk observed a calf 30 m away winded, with heaving flanks and open mouth, gallop to a lake, wade in, drink, then pick its way slowly along the rocky shore. About 5 min later, a wolf came at a fast trot, nose close to the ground, following the calf's trail. From the appearance of the calf, the chase had probably been a long one.

15. September 1952—Rowbrock Lake Area, Northwest Territories, Canada (Kelsall 1957)

A pack cautiously approached a grove of stunted spruce in which about 30 caribou were feeding and resting. When all the caribou were hidden by trees, four wolves spread out along the southeast margin of the grove while the fifth and largest circled around the northeast side, apparently to drive the caribou to the others. This wolf was nearly in position when a bull stood up only a meter in front of one of the waiting wolves, which dashed forward, flushing all the caribou. After a chase of about 200 m, the wolves regrouped and left.

16. July 30, 1953—Brooks Range, Alaska (Crisler 1956)

From 0800 hr to 1200 hr, we watched two wolves hunt. The caribou were stragglers, alone or in small groups drifting in the wake of a large migration weeks ago. The wolves chased a cow going north, both increasing their speeds. The cow drew ahead. The larger, lead wolf, stopped, glanced back at the other wolf and sat. The cow kept running and disappeared. Typical was the wolf's prompt judgment as to when the chase had become futile. Suddenly the smaller, more alert, and determined wolf started back south. (We thought it was a female, and for convenience will call it so.) A young bull caribou was trotting north. The wolf passed him, above, unseen, until behind him, then galloped after him. He ran toward the big wolf, which waited to the north.

As the bull passed below, the wolf cut in on a right angle after him. The bull swerved, crossed a marsh at the end of a lake, then continued on. When the wolf hit the marsh he wallowed and emerged muddy.

The two wolves trotted back south to the lake. Suddenly, the smaller broke into a run and put a big, mature bull into the lake. He started swimming to our side. The wolves trotted down the far side of the lake and headed leisurely along our side. The bull reached shore on our side but stood quietly in the water as the wolves drew nearer. He turned, looking mostly north, toward where he had originally been traveling. Not until the wolves were close did he see them. Instantly, he plunged back into the lake and swam to its northwest corner. The wolves retraced their steps. It looked as if wolves and bull would reach the northwest corner of the lake simultaneously. But as the wolves approached the far side of the lake, they were behind a cow and calf trotting north. The wolves stretched into a run after the oblivious caribou.

The bull neared land just as the two caribou were passing. The three animals startled each other. The bull once more swam toward the center of the lake; the cow and calf fled. The wolves were coming at a dead run. The calf left the cow, turned and ran up the mountainside, where it disappeared. The cow hesitated, often glancing back, even circling. Finally she turned and ran steadily away to the north. She and the calf had given the wolves considerable advantage but had left them easily when pressed.

The wolves lay down. The bull trod water in the center of the lake for quite a while, then swam to our side, remained in the water and stood motionless for a long time. When he did emerge, he stood still another long while, watching before grazing.

Suddenly the female wolf arose and started south. Another caribou was coming north along the west side of the lake. The male sat up and walked sluggishly after the female but kept glancing back. The female, as usual, got behind the cow, then started the run, heading the caribou toward the male. Once the caribou stopped and faced back toward her pursuer. The wolf immediately sat. A wolf prefers not to be eyed when approaching its prey. The cow fled again with the flying upward launch used now and then by frightened caribou. The wolf followed. The caribou had not seen the waiting wolf.

At this moment, as the cow passed near, a calf jumped up from where it must have lain tight all this time. It ran as if to overtake the cow. The cow passed above the big wolf without seeing him. The female wolf had stopped; the chase belonged to the male. He chased the cow but gave up easily. Next he tried for the calf. It acted hesitant, pausing often to stare back at the wolf, which was rapidly gaining on it. Finally, the calf overtook the cow—a feat of fleetness, for the cow was doing her best now, not hesitating at all. Apparently, she was not the calf's mother; again the calf turned and ran up the mountainside disappearing. In spite of its vacillation, this 2-mo-old calf had overtaken an adult caribou by extending itself and had left the wolf behind with ease.

17. April 28, 1954—Brooks Range, Alaska (Crisler 1956)

From 0800 hr to 1600 hr, two wolves approached a low rise and spotted 14 caribou below lying on a frozen "lakelet." The wolves stalked, but without crouching, to within about 30 m before the caribou noticed them and sprang up. The wolves gave chase and could hardly have had a better chance. But the caribou left the wolves behind.

18. June 5, 1954—Brooks Range, Alaska (Crisler 1956)

At noon, four bulls crossed the marshy tundra and sprang into a run. They had aroused two sleeping wolves, which chased them but soon gave up. The bulls paused, then came directly back to where the two wolves stood watching. The wolves gave chase, the caribou split, and the wolves pursued two. The bulls ran along a gravel bar and swam across a river. The wolves ran along the bar, too, but were soon outdistanced and gave up.

19. June 21, 1954—Brooks Range, Alaska (Crisler 1956)

A band of 250 caribou bulls—mostly old, some young—halted on the mountainside and lay down or grazed. Suddenly half banded and ran back toward where they had come. A lone wolf was following them. The caribou traversed the mountainside—its rises, depressions, rough rocks and brush—like a fast-flowing river. Some 200 m behind, the solitary wolf followed. The caribou drew ahead, escaping the wolf.

20. July 12, 1954—Brooks Range, Alaska (Crisler 1956)

Between 1400 hr and 2300 hr, a herd of about 30,000 caribou passed. About 2200 hr, we saw a lost calf searching frantically. Once it rotated in a wheel, bounding up on each turn, to look. Two wolves chased it, but the calf escaped. The wolves then entered a large sector of hundreds of caribou resting or grazing in willows. In the next half hour, the male wolf dropped four calves; the female nosed each kill. They stood over the first kill a few minutes, then ran on, putting the caribou to flight. The wolves ran in the midst of them. The outlying caribou fell in behind them and passed them. The wolves never changed gait; probably they could not run faster.

Two of the calves, not completely dispatched, left their beds and plunged into the river. One we never saw again. The second was found later, partly eaten by wolves, where it had drifted ashore. The wolves returned on the night of the kills and ate the third calf almost entirely. The fourth kill was untouched the next day. Its mother was loitering near it. At our approach she headed to her dead calf as if expecting it to flee. (There is a noticeable return of cows searching for lost calves that have been killed or lost.)

21. July 17, 1954—Brooks Range, Alaska (Crisler 1956)

At 1600 hr, we saw a bull running not 12 m ahead of a lone wolf. The bull jumped into a river and swam across. The wolf did not.

22. June 3, 1957—Keewatin District, Northwest Territories, Canada (Kelsall 1960)

At 2025 hr, three bands of caribou crossed the ice almost 1.6 km away. A few minutes later the herds were running at top speed. A wolf was chasing the largest herd, which was following two smaller groups. The herds numbered 80, 28, and 28. The wolf was 100–150 m behind. Almost 400 m away, a smaller wolf was loping leisurely toward the chase. The caribou changed course and proceeded at top speed. The wolf slowly closed the gap. When the wolf was 50 m or more away, the herds joined in a tight mass. The wolf did not appear to accelerate but kept some 50 m behind.

Eventually an animal in the center of the herd, but toward the rear, tripped, and the wolf spurted and shortened the distance to half before the animal arose and rejoined the fleeing herd. The wolf accelerated and soon was only a few meters behind the hindmost animals. Some caribou veered, but the wolf kept after the one that had fallen and in moments grabbed it by the left hind leg about 11 cm above the hock. The caribou was down immediately. The wolf held it by this grip for 2 min, during which the herd ran almost 1.6 km before slowing. At 2034 hr, the wolf attempted to seize the caribou by the throat, but the caribou managed to arise on its front legs. At 2036 hr, the wolf and caribou were down again. At 2042 hr, the second wolf arrived, and the wolf that had made the kill seemed to be feeding.

23. July 19, 1966—Denali Park, Alaska (Haber 1968)

Three adult black wolves approached eight feeding, cow caribou from downwind. The caribou did not detect the wolves. When the three were 400–500 m from the caribou, they stopped and watched them for 2–3 min. One wolf began stalking low through bushes until gaining a small hill 150 m south of the caribou. A second wolf approached similarly until positioned in bushes to the north. The third remained 400 m away to their west. After 20 min, all three dashed toward the caribou from three sides; it seemed to me that the wolf on the west gave the cue to attack. Although surprised from close range, the caribou all escaped, running tightly between the wolves on their flanks; the wolves gave up.

24. August 23, 1966—Baffin Island, Nunavut, Canada (Clark 1971)

Four wolves were traveling when the leader detected a bull caribou grazing 30 m downslope from their route. The lead wolf stopped with head up, ears forward, looking intently toward the caribou. Instantly, the three wolves following froze until the leader moved slightly crouched behind a rock, whereupon two of the others also moved cautiously behind rocks, and the fourth remained motionless. The four wolves were thus sitting in a small arc behind rocks 30 m upslope from the caribou for 9 min, watching the caribou. Then simultaneously all four rushed the caribou. The caribou escaped downslope, and the wolves milled around for 1.5 min where the caribou had been grazing before resuming their travel.

25. June 22, 1967—Denali Park, Alaska (Haber 1968)

About 300 caribou were scattered in the area, 50–60 of which were calves. A black wolf lay 300–400 m from 22 cows and 10 calves, watching them. When the wolf ran toward them from the northeast, the caribou fled south in a tight, swerving band, calves in the center. As the wolf gained, three cows and two calves veered and were immediately pursued. Shortly, the wolf caught a lagging calf. As the other three cows and a calf continued, the second wolf appeared immediately in front of them and gave chase. Having surprised the caribou from 100 m, this wolf quickly gained and killed the second calf. It seemed that a definite hunting tactic had been used in that the two wolves had positioned themselves on opposite sides of the caribou, with one wolf chasing them toward the other.

26. June 29, 1967—Baffin Island, Nunavut, Canada (Clark 1971)

Three wolves sighted 56 caribou resting 800 m away. Two wolves headed west to the south end of a ridge. The third had disappeared into a small gully and 13 min later was moving cautiously southeast, east of the caribou. The two wolves on the ridge moved slowly toward the caribou, although one soon lay down. When the wolf to the southeast was <400 m east of the nearest caribou,

one started trotting, alerting the others. Instantly, the southeast wolf and the north wolf charged the herd, which split up, some fleeing southeast ahead of one wolf while a group fled west immediately and were pursued by the third wolf. Eventually all the caribou headed west, pursued for varying distances up to 1.6 km by the other two wolves. The chases were unsuccessful.

27. July 10, 1967—Thelon River, Northwest Territories, Canada (Dauphine 1969)

At 1500 hr, a wolf was running 300 m away when a 15–20-d-old caribou calf jumped up 30 m in front of the wolf and perpendicular to the wolf's course. The wolf swerved and bounded to intercept the calf, which ran strongly but did not dodge, even when the wolf drew near. The wolf overtook the calf after approximately 40 m and knocked it down with its shoulder. It seized the calf by the back of the head and shook it violently for 5 sec, dropped it and continued running over a hill. The wolf's teeth had lacerated the 12.5-kg calf's scalp and punctured and fractured the parietal, occipital, and frontal bones of its cranium.

28. August 4, 1967—Denali Park, Alaska (Haber 1977)

Two wolves of the Toklat Pack trotted toward a lone, adult caribou feeding. At about 400 m, one wolf circled to the opposite side of the caribou, where it began a stalk from several hundred meters away. The other approached to within 200–300 m and sat in plain sight of the caribou. The caribou watched this wolf intently, displaying no fear. Meanwhile, the first wolf had stalked through bushes to within 100–200 m, but then the caribou saw it and began running. Both wolves chased but gave up quickly after the caribou widened the gap to several hundred meters.

29. June 30, 1968—Baffin Island, Nunavut, Canada (Clark 1971)

A male and female wolf traveling east stopped, and the female lay down. The male sighted 25 caribou 800 m away. He trotted, then galloped, with increasing speed as the caribou started moving north. Fourteen caribou veered northeast up a small valley while the remaining 11 followed a boulder-strewn streambed northwest, with the wolf pursuing. The wolf gave up shortly after the band split and trotted back to the resting female. Then a yearling caribou fell among boulders and remained down trying to struggle to its feet, which took about 20 min. Meanwhile the wolves headed toward 10 of the caribou the male had just chased. Both wolves then chased them northeast toward where the injured caribou was still standing. The female wolf lay down 90 m from the injured caribou without detecting it. The male gave up 180 m from it, spotted it, walked rapidly in a crouch, then galloped toward it. The caribou managed to go 14 m before the male grabbed it by the right leg. The female then dashed over. The male upset the caribou onto its left side only after the female was pulling on its snout. Within 2 min the caribou had stopped struggling.

30. July 18, 1969—Baffin Island, Nunavut, Canada (Clark 1971)

A single wolf discovered a band of 10 caribou, including two calves, 900 m away and remained in place but with front end raised and head lowered watching them. When the caribou had moved to within 60 m of the wolf and stood watching it, the wolf loped toward them but slowed to a trot somewhat crouched and gained a small circlet of rocks, the tallest of which were a half-meter high. For the next 5.5 min the wolf sat motionless peering over and between rocks at the caribou, which remained curious and hence vacillated between moving toward the rocks and grazing, because they could no longer see the wolf. The caribou moved 90 m away in a tight bunch, but the wolf remained motionless watching. Then the caribou returned, and as they passed the rocks, the wolf rushed them in a fast, crouched trot, then a lope, and finally a full gallop. The chase lasted 4 min, 5 sec, the last 44 sec of which were pursuing a calf that had fallen behind its mother and another cow and had begun zigzagging toward the bottom of a small valley where the wolf caught it.

31. March 1970—Denali Park, Alaska (Haber 1977)

Eight wolves detected an old cow caribou on an open, tundra plateau bordering a river valley, stopped for 1–2 min, and reversed direction until several hundred meters upstream from her. They ascended the plateau, looped around, and fanned out in a broad, concave line toward her. This maneuver forced the caribou to flee down the steep valley through deep, drifted snow and heavy brush. She floundered almost immediately and

tumbled part way down the slope, whereupon the wolves killed her.

32. April 1970—Denali Park Alaska (Haber 1977)

The five wolves spotted 13 caribou from 800 m grazing on a bench and approached steadily from the north where the caribou could not see them. When the wolves reached a location below the caribou, the dominant male left the others and positioned himself 150 m directly north of the caribou still out of view, sat and looked at the other wolves, which had been watching him. When he did this, they moved single file (subordinate male leading) along the bench, until 150 m of the caribou, which could not see them either. These four wolves then looped around, fanned out, and ascended the bench, appearing on top 60 m from the caribou.

The caribou fled with the wolves pursuing more or less line abreast, and the dominant male waiting directly ahead. When the caribou reached the end of the bench they spotted the dominant male below and split into two groups—10 veering one way, and a cow and two calves the other. The dominant male bolted first toward the 10, but in a few seconds reversed toward the cow and calves. The other four wolves had focused on the 10 but also switched to the three. The three caribou gained quickly, and the wolves gave up after 200 m.

33. March 1977—Denali Park, Alaska (Haber 1977)

The 11 wolves approached 10 caribou, which they had detected 800 m away. The caribou saw the wolves from 150 m but did not flee until the wolves were <100 m away and running fully at them. The caribou fled in a tight band on a low ridge forming an 800-m northwest-to-southwest arc (less snow on ridge). Ten wolves fanned out and pursued along this arc, but a subordinate male cut across the arc and met the caribou almost head-on to within 30 m, whereas the other wolves fell behind. After another few hundred meters this male lost ground as well, and in a minute all the wolves gave up.

34. June 19, 1981—Northwest Territories, Canada (Miller et al. 1985)

A lone wolf at 1810 hr was chasing 200–300 caribou. It chased a zigzagging calf for 10 sec, and the calf went down, a male, 4–7 d old, and 9.0 kg. It had tooth punctures or trauma in the upper rib cage and spinal area, right femur, and tip of the right diaphragmatic lobe of the lungs.

At 1815 hr, the wolf again chased the caribou, singled out another calf and came to within 10–15 m of it, but both disappeared behind a ridge. After 15 min the same wolf reappeared chasing a calf along a ridge, killed the calf, picked it up, circled, put the calf down, and then bedded.

At 1833 hr, the wolf arose and easily overtook and killed a third calf.

35. June 17, 1982—Northwest Territories, Canada (Miller et al. 1985)

Several hundred caribou became alert at 1052 hr and galloped from a wolf that appeared. At 1054 hr, amid the confusion following its appearance, the wolf briefly ran and caught a calf by the rump. The calf fell, and the wolf continued without hesitating. In <30 sec, the wolf grabbed another calf by the rump. The calf dropped, and the wolf bit its head before running after more caribou.

At 1055 hr, the wolf left its second kill, about 75 m from the first, and ran upslope after a third calf, which it caught within 15 sec. The calf dropped, but its ears continued to move. The wolf picked it up (seemingly biting it in the head as it did so), then laid it on its side.

At 1056 hr, the wolf ran toward six caribou and disappeared behind a ridge as it chased a cow.

36. September 2, 1986—Denali Park, Alaska (J. Burch)

At 1058 hr, yearling female Wolf 213, 400 m from 26 caribou, trotted toward them and when 200 m away gave a hard chase. The caribou immediately ran hard up hill. After a short distance, the wolf just trotting again, caribou milled about on the hill. The wolf was cautious toward the caribou and started running hard again; the caribou ran but not real hard. The wolf looked a lot slower than the caribou. At 1116 hr, the wolf gave up. The closest wolf got was 100 m.

37. October 28, 1986—Denali Park, Alaska (J. Burch)

At 1535 hr, 7-yr-old (estimated) female Wolf 227 ran up a steep slope, chasing 17 caribou—200–300 m behind. The wolf closed the gap quickly. (The caribou did not seem to be aware that the wolf was closing in.) The wolf came to within one caribou length of the last caribou, but the

animal made it to a plateau, pulled away, then went down a gradual slope, pulling away more. The wolf was still running hard. The caribou continued along the top of the ridge, gaining more; wolf 200–300 m behind. At 1540 hr, 5 min and 300–400 m behind, 227 gave up.

38. January 30, 1987—Denali Park, Alaska (B. Schults)

At 1155 hr, I saw 2-yr-old (estimated) female Wolf 235 and male 237 following a trail 2.4 km from caribou; caribou and wolves became aware of each other about 300 m apart. Wolves gave short chase, and the caribou easily outran them. The closest the wolves got was 300 m.

39. February 26, 1987—Denali Park, Alaska (T. Meier)

At 1806 hr, 4-yr-old (estimated) female Wolf 151 was chasing two antlered bulls plus another caribou 800 m ahead. Observation broken off because of lateness. Wolf still in pursuit.

40. May 14, 1987—Denali Park, Alaska (T. Meier)

At 1015 hr, 4-yr-old (estimated) male Wolf 221 chased six bull caribou, which our aircraft frightened. The wolf gained on the caribou briefly, but they descended to a river, and the wolf broke off at the edge of the gravel bar, getting no closer than 150 m.

41. December 29, 1988—Denali Park, Alaska (A. Blakesley)

At 1338 hr, six wolves (including 349 and 351) were traveling single file. The lead four chased five caribou up a frozen creek but gave up within a minute.

42. May 16, 1989—Denali Park, Alaska (L. D. Mech)

At 1550 hr, yearling female Wolf 361 sneaked to within 10 m of four caribou; they bolted, and she chased around 50 m and gave up.

43. May 17, 1989—Denali Park, Alaska (L. D. Mech)

At 1720 hr, yearling female Wolf 365 sneaked to within 12 m of two caribou adults and a calf all lying and rushed them. The adults ran, and the calf remained flat. The wolf went directly to the calf and killed it. Left it lay but cached hindquarter of previous calf kill, then went on and missed one female calf caribou by 400 m and two adult caribou and calf by 125 m.

44. May 20, 1989—Denali Park, Alaska (T. Meier)
At 1643 hr, traveling yearling female Wolf 365 saw a lone, caribou cow, sat momentarily, chased the caribou <400 m in a semicircle, gave up, and returned to where the caribou had been, as if to look for a calf. She then continued on.

45. May 22, 1989—Denali Park, Alaska (J. Burch)
At 1907 hr, 8-yr-old (estimated) female Wolf 309 started to chase a cow and calf caribou. After a 15–20-sec chase the wolf caught the calf, then just stood and held it while it struggled 1 min, then began feeding. The cow remained 30–40 m off but made no attempt to protect the calf. After the wolf left, the cow approached the calf to within 5–10 m, then ran away.

46. June 29, 1989—Denali Park, Alaska (T. Meier)
At 1111 hr, yearling male Wolf 373 chased a cow and calf caribou, but as soon as the caribou began going downhill, the wolf fell behind and quit.

47. July 11, 1989—Denali Park, Alaska (F. Dean)
At 0655 hr, one medium-sized male caribou was trotting fast down riverbar. One minute later, a wolf trotted fast after it. At 0700 hr, two more wolves trotted down bar. The chase lasted 10–15 min and covered more than 1.2 km. The closest we saw wolves get to the caribou was 500 m.

48. July 21, 1989—Denali Park, Alaska (B. Looney)
At 1700 hr, a bull caribou ran fast across a river bar with a wolf 30 m behind and gaining. Each time the wolf came closer (8 m), the caribou would cross the river. The crossings didn't slow the caribou but drastically slowed the wolf. The caribou ran 500 m up one fork of the river, turned widely, and came back down, without apparent contact with the wolf, and out of sight. A minute later, the caribou and wolf reappeared, disappeared, and reappeared, and the caribou ran up the river fork again, making crossings as the wolf neared.

The caribou was now barely running down the middle of the river while the wolf trotted down the bank. The caribou stopped in the middle of the river, turned, and faced the wolf. The wolf made side-to-side attacks at the caribou, darting toward the right and left side of the head. At each attack the caribou aimed the "shovel" part

of his antlers at the wolf. The whole snout of the caribou was very bloody. After 2 min of very energetic attacks by the wolf, the wolf went a few paces back and lay down panting. The caribou walked a few paces downstream and, with back to the wolf, stood in the stream, tired and motionless with right side facing the wolf, 30 m apart, for 5 min. Two more wolves approached straight toward the standoff, until one wolf broke from the other to arc upstream. Total chase was 3.2 km. From a different vantage point John Burch reported that the wolves were nearly beside where the caribou disappeared from view under a bluff. He was told that it floated out "a few minutes later with one wolf riding the carcass."

49. March 15, 1990—Denali Park, Alaska (J. Burch)
At 1542 hr, a large herd of cow and calf caribou came over a rise and ran into seven resting wolves. The caribou scattered into four or five different groups, with one or two wolves chasing each. The caribou easily outdistanced the wolves. A calf split off by itself down a small ridge; one wolf broke off its chase and targeted it. The calf easily outdistanced the wolf until it tried to cross a gully and floundered in deep snow. The wolf quickly closed the gap (now distantly followed by another wolf), but the calf got through the snow and to a small ridge. The snow did not slow the wolf. When both wolf and calf were on good footing again and the wolf was 10–15 m behind, the calf turned sharply right. The wolf cut the corner and caught the calf. There was no apparent struggle.

50. June 8, 1990—Denali Park, Alaska (F. Dean)
At 0557 hr, a yearling male caribou trotted upstream along a river bar with a wolf slowly trotting 75 m behind. The caribou reversed and ran downstream still on gravel, and the wolf trotted about same speed and lag. This sequence was repeated twice by 0603 hr. The pace of the last pass was a bit faster. At 0630 hr, the caribou was standing in a large channel on a small bar, facing the wolf, which was lying curled up probably 50–75 m from the caribou. When they saw me, both responded nervously—the caribou shuffled a bit but did not leave. The wolf arose and crossed to a vegetated bar where it lay down. The wolf moved a bit and sat and lay down alternately, and it finally continued to a shrubby island. At 0700 hr, I could still see the wolf. At 0732 hr, the wolf walked toward the caribou. At 0749 hr, the caribou stum-

bled to its front knees but got right up. At 0813 hr, I still had not seen the wolf again. The caribou remained in shallow water in the immediate area at least until 1650 hr, but its fate is unknown. At 1049 hr, a double puncture wound was visible on its left hindquarter through the scope.

51. December 7, 1990—Denali Park, Alaska (T. Meier)

At 1223 hr, three wolves were following a well-beaten caribou trail on which hundreds of caribou were moving. The closest group, 120 caribou, walked 800 m ahead of the wolves. The caribou spooked once and ran a short distance. About 2.4 km from where we saw the wolves, the caribou stopped, milled about, and then moved back toward the wolves. The lead wolf went around to the side of the herd, eventually getting behind them. The other two wolves remained on the trail and briefly chased the caribou when they got near. We saw no contact between wolves and caribou. The caribou were separated by the wolves several times, but quickly regrouped and continued on. The smallest wolf, last in line, at one time appeared hesitant when several caribou headed right toward it. The wolves continued along the trail for 2.4 km and passed five caribou 80 m away on a bench covered with deep snow but may not have seen them.

52. August 16, 1991—Denali Park, Alaska (N. Hoffman)

At 0730 hr, a wolf was hunting a single caribou, and after 30 min, a second wolf joined it. The first caught the caribou midstream in a river, biting its neck and back causing it to fall several times until it stayed down.

53. October 24, 1991—Denali Park, Alaska (J. Burch)

At 1308 hr, both Wolf 427 and another began running hard far from six caribou. When they neared the caribou, the caribou easily outdistanced the wolves, which sat and rested or walked slowly, then chased the caribou again. This happened three times. Each time the caribou outdistanced the wolves, then just walked or milled around. The wolves were particularly slowed by following in the caribou's trail when the trail crossed brushy draws.

54. November 18, 1991—Denali Park, Alaska (J. Burch)

At 1345 hr, Wolf 437 plus six others were stalking a large group of caribou. The caribou detected the wolves and ran off in three directions. One wolf ran after them a very short ways, but the rest walked and milled around. The closest the wolves got to the caribou was 800 m.

55. May 25, 1992—Denali Park, Alaska (J. Burch)

At 2030 hr, when we first found Wolves 433, 435, and 459 plus eight others, they were traveling rapidly together— noses to the ground (very fast trotting). Then they detected a caribou cow and calf and started running hard. The caribou curved around in a large area, almost doubling back, which temporarily confused the wolves. The wolves continued following the caribou scent instead of cutting across, but conifers obscured the cow and calf. The wolves slowly closed the gap, saw the cow and calf, and quickly caught the calf.

The other wolves continued to chase the cow, with 2-yr-old female Wolf 435 leading. She ran hard (with other wolves behind), keeping up with the cow. The cow crossed a snow patch. Both lost some ground in the snow. The wolf continued the chase for a short ways past the snow, then gave up.

56. February 28, 2009—Innes-Taylor Rapids on the Little Churchill River, North Manitoba (Kiss et al. 2010)

From a helicopter, we saw a cow barren-ground caribou standing in a small section of open water, motionless and covered with icicles. A gray wolf ran toward it from the nearest treed shoreline. The caribou swam across the river to the opposite side of the ice bank, and the wolf traveled along the ice edge to where the caribou was attempting to climb ashore. As the wolf approached, the caribou retreated into open water. Numerous wolf tracks and much ice attached to the caribou suggested a prolonged chase. When the helicopter disturbed the wolf, it retreated into trees and was undetectable. The caribou climbed ashore and stood motionless for 1 hr until we departed.

On returning, we found the caribou in the water, and the wolf standing along the flow edge. Blood in the snow and an open wound on the right hindquarter of the caribou indicated a struggle. The weakened caribou attempted to exit the water only meters from the wolf. The wolf grabbed the caribou, biting it on the front, left shoulder near the neck. A short struggle ensued, and the caribou died.

Conclusion

The wolf-caribou game of life is in many ways similar to that of the wolf-deer game. Like deer, caribou have evolved a variety of survival adaptations, and wolves have evolved a variety of characteristics to help them exploit any weaknesses in caribou adaptations. Wolf senses of smell, hearing, and vision are especially acute (see the introduction, p. 3), which allows the carnivores to detect caribou much farther than caribou seem to detect them. Thus wolves can approach caribou with maximum opportunity to exploit whatever weaknesses are present in the circumstances in which they find caribou.

The best circumstances for wolves occur during calving season because when wolves do find calves less than 2 wk old, those calves usually just hunker down rather than run. A wolf that finds such a calf nearby grabs it and kills it (account 43). As indicated earlier, caribou tend to space out and space away from wolves to produce their very vulnerable calves. However, when wolves manage to overcome those strategies, the predators can "clean up."

How do wolves overcome caribou-calf survival strategies? One of the wolf's greatest abilities serves it well in this regard—its mobility. Being able to travel long distances for long periods allows wolves to search far and wide for prey (see the introduction, p. 3). Thus even though calving caribou may space many kilometers away from the nearest wolves, the herd may actually only be a few days' travel away. Any given wolf might not know just where to search, but a combination of persistent travel and excellent sensory abilities, especially the sense of smell, would serve a wolf well in locating caribou. Presumably, older wolves would even have the benefit of experience to help. In Canada's Northwest Territories, a female wolf attending a den of pups in mid-July made a 14-d excursion to a herd of caribou some 103 km away (fig. 3.1). Her minimum round-trip distance was 339 km (Frame et al. 2004).

Once wolves find a calving ground, especially if it is during the early stages of calving, they sometimes kill as many calves as they can. Miller et al. (1985) documented a wolf in mid-June killing three calves on a single occasion and possibly four on another (accounts 34 and 35). The average kill rates during these observations were one calf/minute and one calf/8 min, respectively. The

observers found 34 wolf-killed calves in a 3 km² area that apparently had been killed within the previous 24 hr, probably within minutes of each other based on postmortem exams. This find became one of the classical cases of claimed surplus killing (discussed in chap. 9). It is notable that the workers also located two calves that had been stillborn and two that had been born prematurely. This latter observation suggests the possibility that some of the 34 wolf-killed carcasses were also those of markedly inferior individuals.

Although in Denali Park, where neonate caribou calves were radio-collared, Adams et al. (1995, 230) found no difference in birth weight between calves killed by wolves and those that survived, they did note that "the proportion of neonates killed by wolves [in a given year] was more strongly correlated with average birth weight than with wolf density or calving elevation." Mech et al. (1998) cautioned, however, that there are several other immeasurable factors that could predispose newborn calves to wolf predation.

At the same time, pure chance could play a strong role in determining which neonate calves wolves kill, given the calves' weakness and hunkering-down behavior. When Mech watched a wolf in Denali Park merely approach such a hunkering calf and grab it (account 43), it reminded him of watching wolves on Ellesmere Island picking up young arctic hares that were just sitting crouched (chap. 8, "Arctic Hares," account 47). Thus, once a wolf finds such a calf, it may matter little how much the calf weighs or whether it is healthy.

Once calves reach 2–3 wk of age, however, health and condition probably do play much more of a role in which calf wolves kill. Calves then spend most of their time with their mothers and can scamper fast enough to easily outrun humans, although not necessarily wolves. Nevertheless, wolves must work much harder to catch caribou calves as they grow and develop further. As discussed earlier, if a wolf chases a herd of cows and calves, and a calf falls behind, that will be the wolf's target. In account 9, Murie (1944) described just such an encounter. Falling behind "seemed to encourage the wolf to put on added speed, and in less than 400 meters he overtook the calf, knocking it over as he closed in."

Thus, throughout the rest of summer, wolves tend to focus on calves, although the proportion of the total calf crop they kill tends to decrease considerably as the more

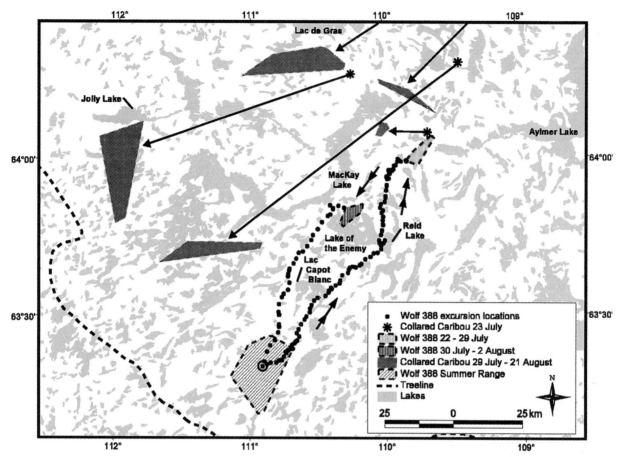

FIGURE 3.1. Record long movement by a denning, radio-collared, female wolf seeking caribou in the Northwest Territories of Canada. The trail of dots illustrates the route of the wolf from the den northeast to the caribou herd and back to the den (Frame et al. 2004).

vulnerable individuals are killed and as the survivors develop and become larger, stronger, and faster (Adams et al. 1995). During this period, the herds of cows and calves wander continually. This behavior forces wolves to spend much time searching for those herds even while they themselves are headquartered at dens with their pups (Mech 1970). By chance, some caribou herds even wander past wolf dens. For example on June 1, 1940, Murie (1944, 169) saw that "all day caribou had been in the vicinity of the den," and on June 19, 1940, he watched about 500 caribou pass within 500 m of a wolf den. In neither of these cases did the wolves attempt an attack, possibly because they tend to hunt prey more on their own terms (see the introduction, p. 8). In fact, wolves from this same den once traveled 16 km to a caribou herd even though during the day 400 caribou had fed within 1–3 km of their den, and two bands of cows and calves had come within 400 m of the den. The general

pattern of daily wolf hunting trips from a den can be seen from the movements of a Denali Park wolf outfitted with a Global Positioning System radio collar in May 1996 (fig. I.2).

Wolves continue to focus on calves even as they reach yearling age and are much harder to catch than when <3 wk old and vulnerability to wolves decreases considerably. Not until caribou reach 2 yr old in Denali (Mech et al. 1998) or 3 yr old in northern Quebec and Labrador (Parker and Luttich 1986) does their vulnerability to wolves decrease. At these ages, caribou begin to reach their prime, so most can outrun wolves. Caribou retain this ability for many more years depending on various circumstances, forcing wolves to continue searching for those members of the herd that cannot outrun them.

Ever the opportunists, wolves try repeatedly to find situations where they can prevail. At a usual tireless travel speed of about 8 km/hr (Mech 1994b) and assuming only

12 hr/d of traveling (summarized by Mech 1992), wolves can cover a huge area and search for many opportunities. South of the tundra, where woodland caribou live, logging roads, seismic lines, and other artificial routes through the wilderness facilitate wolf travel in search of prey, including woodland caribou (James and Stuart-Smith 2000; Latham et al. 2011). When wolves do find herds of caribou, they seem to test them for possibilities. Murie (1944) described one such wolf: "Although galloping hard he did not bear down on any herd. It looked as though he were testing the groups, looking for an especially vulnerable calf. After considerable chasing, the wolf ran after four adults and one calf, driving them off by themselves" (account 8). Murie (1944) also noticed wolves watching caribou herds as they ran away as though looking for calves (account 3). Banfield (1954) reported a wolf that followed a calf trail some 5 min behind the calf (account 14). Meier observed a wolf return to where it had first seen a lone caribou "as if to look for a calf" (account 44). This behavior reminded Mech of similar behavior of wolves approaching locations where they had seen adult arctic hares and then chasing the leverets that had been nursing there (chap. 8, "Arctic Hares," account 5).

As with white-tailed deer, circumstances at the beginning of a chase can be critical. Generally, most prey try to avoid wolves on the hunt when they detect them at a distance. However, when caught short, both caribou and deer often seem to wait until the last minute to flee an approaching wolf or pack. They appear to be saving their energy for a sudden burst of speed that instantly propels them away from their attackers. In many, if not most, cases such a burst discourages wolves from chasing them very far. Such is even true with calves.

If a wolf does finally catch up with a calf, even an older calf, it has no trouble taking the calf down without much struggle (account 49). At about a year of age, caribou weigh only about 40 kg, about the same as a wolf (Kelsall 1968). Thus a single wolf can easily knock one over or pull one down and disable it with a few bites.

During chases of the calves or adults, terrain and topography become important. Those features are where the difference between the long legs of caribou and the shorter stature of a wolf can be critical. Even when both animals are running at the same speed, differences in gait or height can change the advantage of one or the other

animal. As indicated earlier, healthy adult caribou can outrun wolves on flat and even terrain. However, such conditions as deep snow in gullies can hinder either animal. Burch saw a calf in March that "easily outdistanced the wolves" but floundered in deep snow, whereas the snow did not slow the wolf, resulting in capture of the calf (account 49). On another occasion, both a cow and a 2-yr-old female wolf lost ground when crossing through snow (account 55). In still another case, a wolf was slowed by "rough country" (account 2).

When it comes to attacking adult caribou, wolves face a formidable challenge because of the caribou's superior speed. As indicated, it is the variation in circumstances (terrain, topography, weather, and caribou vulnerability) that helps wolves overcome the caribou's advantages. In addition, wolves might employ certain strategies that improve their chances, as discussed in this book's introduction.

In any case, it seems wolves rarely get a chance to make an easy caribou kill except during calving season or especially deep snows (Mech et al. 1998). Regardless of any use of strategy, catching a caribou generally takes numerous attempts (table 3.1) and considerable effort. According to Murie (1944, 164), "It seems that wolves run adult caribou until they are exhausted, for in a short chase the grown animals can run away from the wolves." Crisler (1956, 341) stated that "the longest chase we saw a wolf make was about five miles [8 km]. There was no kill." Most of the time when wolves chase caribou it seems that, unless unusual circumstances prevail or the chase covers unusual terrain, the wolves seldom come close to the caribou. By way of contrast, sometimes they do come within a few meters but still fail to catch the animal (account 37). Crisler (1956, 339) was impressed with "the wolf's prompt judgment as to when the chase had become futile."

If wolves fail to capture caribou during most attempts, what are the circumstances under which they finally succeed? The answer, of course, is that sooner or later wolves find disadvantaged caribou. As with most other prey, caribou that fall prey to wolves are usually the young and the old, those in poor nutritional condition, or those with arthritis or other physical impairments (Murie 1944; Crisler 1956; Miller 1975; Parker and Luttich 1986; Mech et al. 1998).

Individual caribou in poor condition often stand out

in the herd or, when pressed by wolves, automatically reveal their impairment.

> The caribou the wolf catches is usually the caribou that slows down. Most of the caribou we have seen wolves chasing were healthy enough to run away. In every chase we have seen, it was apparent within a minute or two after the actual chase began whether a caribou would be caught or not. Usually, as I have said, a band swerves and runs as if it were one object with many legs. But a caribou that is subsequently caught loses the speed of the band almost immediately. Sometimes it takes its own direction, not trying to follow the band. (Crisler 1956, 342).

Murie (1944, 176) had also noticed the same thing: "Crippled animals were noted on several occasions. Often they brought up the rear of a band." Murie then described 14 such caribou he had seen, including "the front leg of a cow was bent as though it had been broken and had later healed" and "bringing up the rear of a band of 800 caribou I saw a calf limping on a front foot, a thin cow limping on a front foot so that her head went up and down violently with each step, and another cow with a hind leg which most of the time was not used." In our own studies, in Denali we found three wolf-killed bull caribou during February whose marrow fat averaged as low as 33% and a wolf-killed cow as old as 22 yr; of 66 wolf-killed bulls, 13% had arthritis and 5% had jaw necrosis (Mech et al. 1998).

During periods of deep snow (at least 100 cm), wolves can also cash in on caribou by killing several in a single hunt. In Denali, we found several instances of wolves making two to 17 kills in one hunt during the winters of 1989–90 and 1990–91 (Mech et al. 1998). The average ages of these caribou was younger than those taken as individuals during periods of less snow, but their average marrow fat was 66%, relatively low (Mech and Del-Giudice 1985).

For most of the year and during most winters, wolves must invest much more time and effort to catch caribou. When they do find and catch vulnerable caribou, however, they have no trouble killing them. They can usually dispatch a calf with a single bite or two, especially in the head, spine, chest, or throat. With adults, wolves grab them anywhere they can, including a hind leg just to stop them (account 22) and then grab them more toward the head or neck. It is hard to know just where the killing bite is made on any prey animal because once the quarry is down, the wolves start eating, and the moment of actual death gets obscured. Regardless of where the killing bite is made, the wolf has succeeded one more time in overcoming a caribou's many defenses. Unless the animal is a calf, it most likely is an old adult whose defenses have already served it well for many years and allowed it to produce many offspring. In this way, both the caribou as a prey species and the wolf as a predator continue to survive and interact over the millennia.

4

Elk

ELK ARE ARGUABLY the perfect wolf prey: medium sized so as to not be too formidable, like moose or bison, but large enough to provide a good meal for a pack. About half the elk that wolves encounter flee when attacked, with wolves giving chase (fig. 4.1), while the other half stand their ground (fig. 4.2; MacNulty 2002). This mixture of strategies makes sense in that elk share attributes of smaller prey that usually flee (deer, caribou) and large prey that usually stand (bison, musk ox). In short, elk constitute a compromise between speed and strength. But they still provide substantial biomass, as adult cows weigh about 225 kg, whereas bulls weigh about 350 kg (Peek 1982). Nevertheless, wolves kill elk less than 25% of the time, depending on pack size and other factors (Smith et al. 2000; MacNulty 2002; MacNulty et al.

2012). Additionally, elk occasionally kill attacking wolves (MacNulty 2002; Yellowstone National Park [YNP] Wolf Project, unpublished data). So are elk really such perfect prey for wolves?

Currently, elk are not common wolf prey in many areas of North America although their European counterpart, red deer, seem to be taken preferentially in Europe (Peterson and Ciucci 2003; Pilot et al. 2006). Once widely distributed roughly across mid–North America, elk were quickly reduced on European settlement and mostly limited to the Rocky Mountains (Peek 1982). Wolf eradication was even more extensive.

Some early researchers examined wolf-elk interactions anecdotally (Cowan 1947), but little focused study was done until the late 1960s. Carbyn (1974) examined

FIGURE 4.1. The Leopold Pack chases a bull elk across Blacktail Deer Plateau in Yellowstone National Park, 2007. This is typical behavior for elk when approached by wolves, and wolves prefer to attack fleeing elk because standing elk are more dangerous. (Photo by Matthew Metz.)

FIGURE 4.2. A much more difficult situation for wolf attack. Standing prey are better able to defend themselves, lashing out with front legs and even antlers; wolves are wary and reluctant to attack such prey. Leopold Pack in Yellowstone National Park, 2007. (Photo by Matthew Metz.)

FIGURE 4.3. Classic attack point—on the hind leg and not hamstringing, a common misperception about wolf attacks. Blacktail Pack on the Blacktail Deer Plateau, Yellowstone National Park, 2012. Dominant female Wolf 693 grabbing the bull elk hind leg. (Photo by Rebecca Raymond.)

wolf predation on elk in Jasper National Park, Alberta. His documentation of wolf-elk encounters emphasized their variable nature (accounts 1 and 2). The attack points were typically from the rear (legs and sides), with opportunistic grabbing of the neck, later confirmed in other studies (see Landis footage [1998] and figs.4.3–4.5).

A series of studies in Banff National Park in Canada examined wolf-elk ecology (Huggard 1993a, 1993b; Hebblewhite 2000), with Weaver (1994, ii) concluding that elk were especially important to wolves, calling them "the most profitable prey with minimal risk-of-injury." Pilot et al. (2006) found that European wolves killed elk preferentially over other prey, as had Weaver (1994) with elk in Canada. These conclusions are consistent with the "perfect prey" idea. Weaver's tracking of wolves suggested a mode of regular travel across a territory searching for prey that they knew were there. Their route could take from a few days to a few weeks (1994, 77).

Weaver (1994), Hebblewhite and Pletscher (2002), and others also stressed the importance of elk grouping as defense against wolves (introduction and chap. 1). Grouping diluted the probability of an individual being killed and sometimes reduced detection probability because groups were unevenly distributed (www.press.uchicago.edu/sites/wolves, video 8; fig. 4.6). In essence, grouping creates large holes on the landscape where there are no prey, more advantageous to elk than being sprinkled across the landscape where encounter rates would be higher. This actually drives down consumption rates (Fryxell et al. 2007). Hebblewhite and Pletscher (2002) and White et al. (2012) found groups of six to 30 elk to be the herd size in which elk were the least safe from wolves, but herds of that size were scarcer, the scarcity being a possible second type of defense against wolf predation.

Elk are also critically important to the endangered Mexican wolf. Only about 83 of these smaller wolves exist in the wild, and they prey more on elk than on deer, despite deer being more available (Reed et al. 2006). Group hunting might allow Mexican wolves to overcome any size disadvantage (MacNulty et al. 2012).

Elk grouping behavior is complex. For example, when wolves were restored to Yellowstone National Park in 1995, elk group size in the interior increased from an average of 5.6 (SE = 0.17) to 8.7 (SE = 0.17), and energy requirements, habitat, landscape, and predator commu-

FIGURE 4.4. Six Druid Peak Pack females chasing a bull elk in Lamar Valley, Yellowstone National Park, 2007, assessing capture opportunities. The two large males (302 and 480) in this pack did not participate in the hunt for unknown reasons, but the other wolves pursued the elk until he stood his ground, where the female wolves immediately stopped and retreated without further attack. Usually, this would have been the point where the larger males might have pressed the attack and gotten the tired elk running again (the females had chased it for approximately 3 km) leading to a kill. In this case, the elk went unscathed. (Photos by D. W. Smith.)

FIGURE 4.5. With smaller prey, wolves typically grab the neck, as the risk of injury to the wolf is less. However sometimes they also grab the neck on larger animals, such as this bull elk, as it is very effective at rendering prey immobile and safe to consume. Blacktail Pack on Blacktail Deer Plateau, Yellowstone National Park, 2009. (Photo by D. W. Smith.)

nities all played a role in determining group size (Gower et al. 2009a).

Besides adjusting group size, separating themselves from wolves through migration is another important elk defense. Wolves generally den in April and, in mountainous Yellowstone, at low elevations where snow is least. Elk calve in late May or early June, after much snow melt, providing access to higher elevations away from wolf dens. To hunt elk, wolves must travel far from their dens, making hunting time-consuming and energetically expensive (Mao et al. 2005). In northern Yellowstone, cow elk that winter at low elevations migrate immediately to high areas when snow begins to melt (White et al. 2010). Cow elk also separate from each other when they calve, unlike the rest of the year, making it harder for wolves to find a hidden birthing spot (Barber-Meyer et al. 2008).

In winter, elk must return to low elevations due to deep high-country snow. Then, elk and wolves are con-

centrated and share a greatly restricted area (Mao et al. 2005). Winter snow is important to wolf-ungulate relationships (fig. 4.7). For example, deep snow causes virtually all mule deer and pronghorn antelope to leave Yellowstone and also significantly influences bison movements (Geremia et al. 2011). However, for some elk, avoidance of potential predation is so strong they choose deep snow over wolves. Nineteen cows that historically occupied the Madison and Firehole River drainages in Yellowstone migrated to new deep-snow locations, presumably to avoid reintroduced wolves (Gower et al. 2009b).

Further, it appears that wolves themselves limit access to elk in winter due to territoriality. Wolf density is so high in northern Yellowstone that territories overlap (Smith and Bangs 2009). Thus, even with heavy wolf predation, elk still choose the best habitat, consistently in the same areas, winter to winter (Kauffmann et al. 2007). Wolves attempt to occupy these areas, but ter-

FIGURE 4.6. Grouping is a common strategy to minimize chances of being killed, as it reduces an individual's likelihood of attack. Blacktail Pack on Blacktail Deer Plateau, Yellowstone National Park, 2009. (Photo by D. W. Smith.)

ritorial defense limits the number that can. Thus, they are forced through competition to share the elk. This results in wolf-pack territories not centered on prime elk wintering areas (at least as determined by radio-collared cow elk). Elk, then, often frequent wolf-pack territorial edges, which increases the likelihood of wolf packs encountering each other, leading to interference—similar to wolf and deer spacing (Mech 1977a, 1977b). Occasionally in Yellowstone, three wolf packs feed on an elk kill, suggesting intense competition (YNP Wolf Project, unpublished data). If wolves had all the elk in the center of their territory and did not have to compete with neighbors, kill rates might be higher (Mech 1977a, 1977b). Indeed, the leading cause of wolf death (excluding disease to pups) in Yellowstone is other wolves (Smith and Bangs 2009; Cubaynes et al. 2014).

Winter still takes its toll on elk. One cannot overemphasize the effect of deep snow on wolf-elk interactions (Mech et al. 2001) and its annual variability, which seems to be increasing with climate change (Dai 2011). Some winters are warm and bring little snow, while others are snowy and cold. Some research indicates the former are increasing (Rehfeldt et al. 2012). Over the last 20 yr in Yellowstone, the approximate time since wolf reintroduction, two winters stand out as severe, with most being mild. In fact, from 1998 to 2008 every winter was mild with below-average snow depth (Smith et al. 2009). During both severe winters (1996–97 and 2010–11), wolf kill rates were elevated, with some surplus killing (Mech et al. 2001; Smith et al. 2009). In March 1997 the Druid Peak Pack in Lamar Valley killed five elk in 1 d (Mech et al. 2001); this 1-d record was due to winter severity increasing elk vulnerability. The wolves (and scavengers) consumed all five of the elk in 2 wk. This elucidates interesting wolf behavior—they kill as much as they can when killing is easy because such opportunities are rare, and if undisturbed, such surplus is entirely consumed (chap. 9).

FIGURE 4.7. Severe winters with deep snow constitute a major vulnerability for elk and other ungulates, usually making the prey easier for wolves to kill. Slough Creek Pack, Yellowstone National Park, 2008. (Photo by Daniel Stahler.)

Wolf-elk summer interactions are different. With vastly more area available, wolves must search more extensively to find elk, and when they do find them, elk practice several strategies to avoid becoming wolf scat. Like other types of deer, elk often escape to water. Several of the hunting accounts and film sequences by Bob Landis (1998) nicely show this tendency. Coauthor D. W. Smith (unpublished data) observed Wolves 48F, 70M, 72M, and 92M from the Nez Perce Pack chasing Yellowstone cow elk, which ran to Beach Lake, swam across, and continued fleeing. At the shoreline, the wolves hesitated and gave up. If the water is elk-chest deep or the current swift, elk typically escape (www.press.uchicago.edu/sites/wolves, video 9; figs. 4.8, 4.9). If the water is not deep or the current not swift, wolves may actually benefit, almost making it seem like a mistake for the elk to run there, as it is common in Yellowstone to find wolf-killed elk in or near water (Kauffman et al. 2007; YNP Wolf Project, unpublished data).

Coauthor D. W. Smith observed an encounter in Pelican Creek where these circumstances were exemplified (figs. 4.10–4.13). A bull elk fled to Pelican Creek where the water was shallow. The bull ran up and down the shallows hesitating to exit while the wolves followed, and they eventually killed him (account 13). Even in winter, elk may still run to water. One bull ran to a partially frozen pond and broke through the ice, considerably favoring the Swan Lake Pack of wolves (fig. 4.14). Similarly in interior Yellowstone, elk frequently escaped to the Madison River, a tactic that might have allowed elk numbers to remain higher than otherwise (Gower et al. 2009a).

Another twist to the wolf-elk summer story is elk using certain habitats to their advantage. In 1988, 36% of Yellowstone burned, much of it forest (Wallace 2004). Over the years, these dead trees have toppled, creating a morass of tangled logs difficult to travel through. Elk preferentially use these areas in summer, probably because of the ease with which they can move through the

FIGURE 4.8. Mollie's Pack with an elk stranded in deep, swift water. Each time a wolf attempted to approach this elk, the wolf was swept away, and the elk survived. Mollie's Pack in Pelican Creek, Yellowstone National Park, 2003. (Photo by D. W. Smith.)

FIGURE 4.9. Bechler Pack wolves in 2007 attempting to capture an elk from a group of two cows and a calf that have retreated to water. Refuge in water is a common strategy for elk in Yellowstone National Park. A disproportionate numbers of kills are found near water. This hunt was unsuccessful. (Photo by Matthew Metz.)

FIGURESS 4.10–4.13. Elk retreat to water upon wolf attack even when the water seems too shallow to offer an advantage, as this bull elk in Pelican Creek, Yellowstone National Park in 2006 demonstrates. He ran back and forth in the same stretch of stream while the wolves attacked continually, eventually killing it. Attack points were at the rear. (Photos by D. W. Smith.)

FIGURE 4.14. Another example of when running to water is a poor strategy. Why elk flee to shallow or partly frozen water is unknown. Three-year-old (prime-aged), dominant male Wolf 295 of the Swan Lake Pack (lighter wolf) led this attack with help from another prime-aged male. The 7-yr-old, dominant female Wolf 152 stood on the shoreline and watched the attack. Yellowstone National Park, 2005. (Photo by D. W. Smith.)

tangles compared to the shorter-legged wolf (Mao et al. 2005; Forester et al. 2007). It is amazing to watch, while horseback riding through such terrain, how easily a harem of elk slip through this forest rubble—horses balk at such a feat. Wolves undoubtedly have difficulty as well (figs. 4.15 and 4.16).

Unburned forests are also important to elk evasion of wolves. Where wolves are present, elk prefer forest habitats (Creel et al. 2005). However, in northern Yellowstone, there is so much more open terrain (Mao et al. 2005). Risky and safe habitats may vary by location.

Eventually, despite all the strategies to avoid wolves, elk still will encounter them. In the Greater Yellowstone Area, wolves came within a kilometer of each elk an average of three times per month (Middleton et al. 2013). Often the elk flee, the wolves give chase, and most often the elk outrun the wolves (MacNulty et al. 2007, 2012). Speeds for wolves (Stenlund 1955; Mech 1970) and elk

(Link 2004) are approximately the same (56 km/hr). Elk stand their ground about 55% of the time when wolves approach, and wolves rarely kill those (MacNulty 2002; www.press.uchicago.edu/sites/wolves, video 10). Then why don't elk always stand? One answer is because they do so well fleeing that it is not selected against. Given an elk's intermediate size, both strategies work, but elk might be slightly better adapted to fleeing. Indeed, the larger bulls are more likely than cows to stand. Typically when wolves are killed by elk or other prey (10 instances in Yellowstone), it is when the prey are standing and defying the wolves (YNP Wolf Project, unpublished data; fig. 4.17). Such examples attest to the risk wolves incur to make a kill and highlight why wolf hunting strategy is best described as risk averse (www.press.uchicago.edu/sites/wolves, video 11).

Once the chase has begun, landscape features become important. Several times Coauthor D. W. Smith has

FIGURES 4.15 AND 4.16. Post 1988 forest-fire evidence suggests that elk preferentially seek out burned forest, possibly because the downed trees present more difficulty to the pursuing wolves than to the fleeing elk. Leopold Pack, Yellowstone National Park, 2007. (Photos by Matthew Metz.)

FIGURE 4.17. Standing elk present problems. The wolves chased this elk, and then it stood its ground defying the wolves. The Leopold Pack in Yellowstone involved in this hunt were primarily female. In the middle is a 3–4-yr-old female, Wolf 592, and the wolf in front is yearling female 593, but the dominant male of the pack, Wolf 534, was also involved. The 7-yr-old, dominant female Wolf 209 was present but participated very little in the hunt, which was unsuccessful. (Photo by D. W. Smith.)

aerially observed elk easily outdistancing wolves on flat, open terrain free of obstacles. As the chases advanced, however, the elk encountered uneven terrain or a gully or a stream, usually choked with debris like downed trees, rocks, or even thick bushes that slowed the fleeing elk. The downward slope through obstacles allowed the wolves to gain and sometimes make a kill.

These observations reflect the "landscape of fear" concept described by several authors (Brown et al. 1999; Laundre et al. 2001). Elk may avoid certain landscape features because the susceptibility to wolves is greater. Such risk factors may be terrain, firm versus soft footing (Carbyn 2003), or visibility, including the height of willows, which could obscure approaching wolves (Ripple and Beschta 2004). The degree to which such risk factors actually influence elk behavior, however, has been questioned (Middleton et al. 2013).

Understanding elk response to wolves is complicated.

For example, elk often walk right by wolf dens, areas one might think elk would avoid at all costs (also see chap. 1, which indicates that deer also do not avoid wolf dens). Studying elk behavioral response to wolves is difficult because it requires following elk around documenting their movements on a very fine scale comparing behaviors with and without wolves. Global Positioning System collars will help solve this problem. Other questions relate to how wolves hunt. Wolves often hunt during the night, twilight hours, or dim light (Mech 1966a, 1992; MacNulty 2002), so visual cues may not be as important. Although elk vision is not keen, elk olfaction and hearing are, and these are influenced by landscape (Hudson et al. 2002).

It does make sense that elk would adapt behaviorally to living with wolves after years of not, for their lives depend on it. For example, wolf activity parallels that of elk—evening, night, and morning—making it likely that elk adjust to this pressure (Hudson et al. 2002). The documented differences in distribution, grouping, migration, and habitat preferences attest to that. We just have not yet documented the subtler behavioral changes. Byers (1997), a pronghorn antelope researcher, calls ridiculously fast pronghorn (pronghorn run about 30 km/hr faster than wolves) the "ghost of predators past," as their high rate of speed has been evolutionarily maintained despite no predator running that fast. Predator and prey evolved together over millennia, so these behaviors are genetically coded and not easily lost. Moose quickly adjust when predators return (Berger et al. 2001).

Another aspect of elk adaptation to wolves is vigilance. Elk vigilance is a function of wolf-pack size, distance and recency of kill, sex/age and size of elk herd, distance to forest, depth of snow, and season (Liley and Creel 2008). Interestingly, elk in denser wolf populations scanned for disturbances less than elk in areas of lower wolf densities (Creel et al. 2008). Vigilance increased with herd size up to 20 elk then dropped as predicted by the "many-eyes" hypothesis (Liley and Creel 2008). However, other researchers found that elk vigilance did not change before and after wolf reintroduction, suggesting insensitivity to greater predation risk, probably because elk eat and watch at the same time—or chew and scan simultaneously (Gower et al 2009b). The design of these two types of studies differed—one compared elk vigilance across wolf densities, while the other compared elk vigilance before and after wolf recovery.

Finally, some research suggests that wolves reduce elk pregnancy rate by altering elk habitat selection in favor of safer but nutritionally impoverished habitats (Creel et al. 2009; Christianson and Creel 2010). Elk need approximately 10% body fat in fall to conceive successfully (Cook et al. 2004). Thus, wolf-caused changes in elk habitat use (e.g., shifting from preferred grassland to forests [Creel et al. 2005]) could plausibly minimize fat deposition, thereby reducing pregnancy rate. This mechanism has been invoked to explain the apparently negative relationship between wolf densities and fecal progesterone levels (Creel et al. 2007) under the assumption that such levels are an accurate gauge of elk pregnancy rate (Creel et al. 2007, 2009). But this explanation is problematic for several reasons. First, the fecal progesterone assay is an unreliable measure of pregnancy rates because it is prone to producing false

negatives (Garrott et al. 2009). Second, numerous studies using more reliable methods indicate that elk pregnancy rate is unaffected by wolf predation pressure as indexed by wolf numbers or elk-wolf ratios (Hamlin et al. 2009; White et al. 2011; Middleton et al. 2013; Proffitt et al. 2014). Finally, the northern range of Yellowstone supports one of the highest densities of wolves in North America (Smith and Bangs 2009), yet average pregnancy rate (determined via serum assay of pregnancy-specific protein B) for prime-age elk (4–9 yr old) has remained high at 87% (White et al. 2011) and for elk 7–12 yr old, >90% (YNP, unpublished data). Lastly, Boonstra (2012) rejected the Creel et al. (2009) hypothesis for the above reasons and argued, in addition, that wolf-induced stress on elk is acute and not chronic and is therefore adaptive. Thus until more studies support that finding, it should be treated cautiously.

Studies of wolves and elk have greatly expanded recently due to wolf recovery in the northern Rocky Mountains and specifically because of detailed long-term, observational studies in Yellowstone (Vucetich et al. 2005; Smith and Bangs 2009; MacNulty et al. 2009a, 2009b, 2012; Vucetich et al. 2011; Metz et al. 2012). One thing is clear: elk are key to wolves wherever the two co-occur, but there is still much to be learned about their interactions compared to other wolf-prey research.

Hunting Accounts

Prior to research on wolf-elk relationships in Yellowstone National Park, few researchers had observed wolf-elk hunts. Since wolf restoration (Bangs and Fritts 1996), over 500 hunts have been observed. Bob Landis (1998) has filmed many of them (see this footage at www.press.uchicago.edu/sites/wolves), and park naturalist Rick McIntyre has personally observed 92 hunts. Therefore, only a tiny selection of the Yellowstone hunts are reported, while virtually all of the non-Yellowstone hunts are included because they are so few. These selections do not reflect wolf hunting success, as most observers had biased reporting toward successful hunts. Rather, the accounts were chosen to exemplify the variety of wolf attacks on elk and enhance understanding of them. As indicated earlier, wolf success rates hunting elk fall in about the same general magnitude as wolves hunting most other ungulates except bison and moose.

1. June 23, 1970—Jasper Park, Alberta, Canada (Carbyn 1974)

At 0855 hr, an elk was uttering a high-pitched alarm call. A cow and calf were being rushed by six wolves in an open meadow, and more wolves were believed to be nearby in adjacent cover. Five wolves (three presumed to be yearlings) rushed the cow to separate it from the calf; then the order was quickly reversed, and the wolves converged on the calf. One wolf kept the cow at bay. One adult and three yearlings attacked the calf's flanks while another animal leapt on the back biting into the neck. The whole sequence took about 20 sec. The three presumed yearlings remained with the calf while the three larger wolves (adults?) attacked the cow. The sequence was then interrupted by the presence of a human.

The wolves hesitated and slowly left except for one, which remained at the carcass until the human approached. The cow returned at 0920 hr, prodded the carcass with front feet, and nudged (sniffed?) it. At 0937 hr, the cow continued grazing but left when hikers passed through. Return visits of cow to dead calf were recorded the next day at 2020 hr, 2145 hr, and 0815 hr.

2. May 27, 1972—Jasper Park, Alberta, Canada (Carbyn 1974)

A single wolf was seen in a meadow at 0802 hr approaching a single cow elk in a "casual" manner. The cow walked to the opposite side of the meadow holding her head high in alarm. The wolf lay down at 0807 hr, and the cow stopped. Then, slowly, the cow approached the wolf to within 23 m, and the wolf retreated. The cow continued to follow, and at 0809 hr she charged the wolf, which had reached the edge of the trees. The cow resumed grazing and at 0817 hr was approached by a second cow. The second cow sensed the wolf and slowly retreated to the opposite end of the meadow. The wolf returned to the center of the meadow at 0827 hr, and the first cow left at 0837 hr. The second cow was less aggressive and maintained a considerable distance from the wolf. At 0900 hr, 13 cow elk entered the meadow and moved to a lick where the wolf stood. The wolf crouched and watched the herd approach. The cows were running, and individuals exhibited sporadic outbursts of playful (aggressive?) behavior. About 47 m from the wolf, the herd stopped and gradually retreated toward from where she had come. Though alert, the elk did not seem alarmed.

At 0917 hr, the wolf left, and the cows continued to graze in the meadow.

3. April 7, 1996—Yellowstone Park, Wyoming (D. MacNulty)

At 1820 hr, five black pups (almost 1 yr old) and one gray wolf of the Rose Creek Pack were bedded, and at 1822 hr, two blacks trotted downhill toward 40+ elk. The first wolf sped up, and the elk bunched and ran downhill. The second wolf sat and watched. The pursuing wolf pushed about 15 elk uphill, circled back uphill and targeted three cows and a bull. A cow faced the wolf with head raised and stamping the ground. The wolf lunged tentatively at the cow, which then struck at the wolf with raised forelimbs and ran the wolf off. The cow then rejoined the herd.

At 1832 hr, female Wolf 17 left the bedded pack and trotted off. At 1835 hr, she was sprinting among a herd of 75+ elk with two black pups and Wolf 8. At 1837 hr, the herd panicked and split into several directions, and Wolf 17 was suddenly seen tearing at a fresh kill.

At 1839 hr, three black pups and Wolves 8 and 17 started uphill and surprised a cow elk. Wolf 8 locked onto the cow's nose and held on while 17 grabbed her neck. The cow kicked up her forelimbs, and 17 retreated but regrasped the cow's neck. At 1843 the cow fell, and 17 and a black pup tore at her midsection.

4. November 11, 1996—Yellowstone Park, Wyoming (R. McIntyre)

At 1654 hr, I spotted the five Leopold Pack wolves to the west, breeding male 2, breeding female 7, and all three pups. At 1738 hr, Wolves 2 and 7 chased an adult cow elk out from behind a ridge, Wolf 2 leading. He caught up with the cow, leapt up, and grabbed her right hind leg, just above the hoof. She kicked back, but he held on. The cow was slowed by the drag of the wolf. Wolf 7 ran in, leapt up, and grabbed the cow's throat. A gray pup ran to the site, and a black pup ran up and bit the side of the elk. The breeders and the black pup pulled her down.

5. March 3, 2001—Yellowstone Park, Wyoming (R. McIntyre)

Druid Peak breeding Male 21 and a gray pup looked at a nearby elk herd to the southwest. Wolf 21 moved toward the herd.

Another gray pup sat close to those elk and watched them at 0720 hr. The elk moved toward that pup and a black pup behind it. Wolf 21 and three other pups were just downhill to the southeast. The wolves ran at the elk at 0721 hr. A gray pup was leading. Another gray pup was second, and 21 was third. Wolf 21 passed the second pup. The lead gray pup and 21 chased the main herd, heading west, and both wolves disappeared over the ridge.

A cow standing apart from the herd did not run when the wolves chased the herd, and a gray pup approached her. She tried to run downhill but stumbled in the snow. When she arose the pup grabbed a hind leg and easily pulled her down. The cow arose again, kicked her hind legs, shook the pup off, and ran downhill. Four gray pups and two black pups chased her. She tripped in deep snow and slid down a steep slope and rolled over. As she arose, a gray pup ran up and bit her rear end. A second gray pup came in, but the cow was now running downhill.

The cow fell again. Not sure if she tripped or if the pups pulled her down. Three gray pups and a black pup were on top of her, biting her. She was unable to arise. Another black pup and another gray joined the attack. At 0723 hr, Wolf 21 and a black pup ran downhill and joined the six pups with the dead cow.

6. June 1, 2001—Yellowstone Park, Wyoming (R. McIntyre)

At 2110 hr, I saw 14 Druid wolves moving west, 10 blacks and four grays. Three noncollared black yearlings led. Breeding male 21 was seventh in line, and breeding female 42, 12th. Around 2111 hr, the lead yearlings started running west like they were pursuing something. Three small bull elk were running from them. The wolves ignored them and veered south, then ran west after a big bull that had approached a black yearling earlier in the evening.

When the bull swung south, a black wolf cut him off and he continued west. Two black yearlings caught up with the bull and easily ran alongside him. The bull was running slowly. Two more black yearlings ran beside him, and three grays and two additional blacks joined the four wolves with him. All nine were running next to him but none contacted the bull yet. Wolf 42 and another joined the others a minute later. Wolf 21 and three other wolves were somewhere behind the bull at this point.

The bull swung north-northwest and to a small group of bulls. The wolves stayed with him as the other bulls veered off. One wolf right behind the bull suddenly got knocked backward, presumably kicked by one of the bull's hind legs. A black yearling ran a meter in front of the bull, with 106 at the bull's side.

The bull kicked out at the wolves with hind legs while running north toward a river, probably trying to escape to water. Wolf 106 nipped at the bull's abdomen as she ran alongside him. The bull slowed, and 14 wolves attacked him. Two black yearlings leapt up to bite his throat but could not hold on. The bull collapsed but got right up. The wolves swarmed around him. At 2115 hr, Wolf 21 leapt up and grabbed the upper part of the bull's throat. A yearling bit his rear end, and other wolves attacked his sides. The bull dropped again, briefly arose, then collapsed once more. Wolf 21 maintained his holding bite on the upper throat. The bull struggled to stand, but the wolves held him down. He seemed dead at 2116 hr.

7. March 3, 2002—Yellowstone Park, Wyoming (R. McIntyre)

At 0958 hr, 14 Druid wolves chased an elk herd northeast of them. A gray pup led, and a black and 217 were right behind. They soon gave up, and the herd ran farther northeast, stopped, and looked back at the wolves.

A big bull that had been with the herd veered off and moved southwest toward some trees. The lead wolves had run by him while chasing the herd. He stood alone south of the wolves. A gray pup and a black left the other wolves and walked uphill toward the bull. He took few steps away, and the two wolves romped playfully after him. The bull stopped and faced them. The wolves backed off. A second black approached the bull. Breeding female 42 and three other blacks were bedded just downhill from the bull. They glanced toward him but did not seem interested in him.

The two blacks and the gray pup surrounded the bull, which was standing his ground. When he turned away from them, the three wolves walked toward him. He turned back and took a few steps toward them, and they backed off. A third black arrived but soon wandered off. The bull walked off to the southeast. A second gray joined the three wolves near the bull, and the wolves followed. He soon stopped and faced them. The wolves were very close and moving closer. The bull held his head high and he turned from one wolf to another. Then he slowly moved downhill, and the four wolves followed. The bull started running downslope but quickly turned to face the wolves. One of the three blacks with 42 arose, looked at the bull, then walked uphill toward him, joining the two blacks and the two grays. A black moved in front of the bull and romped back and forth. A second black with 42 arose and watched the interaction.

The bull stood his ground against the nearby wolves but suddenly ran downhill. The three blacks and two grays with him chased him. Wolf 42 stood up to look at the nearby chase just east of her. Wolf 42 and a black ran after the bull, but she soon stopped and just watched the others chase him. A third gray joined the chase. There were now four blacks and three grays chasing the bull downhill.

The bull fell in deep snow, and the wolves were immediately on him, one biting his rear end. When he arose, another wolf was holding a hind leg. The wolves pulled him down again. Wolf 42 ran to the site. One wolf was biting the bull's neck. More wolves ran in. There were 12 around the bull. He arose again. Wolves bit his side and hind legs. The wolf at his neck was no longer there. The bull tried to run downhill but was slowed by wolves hanging onto his hind legs. He dragged them along with him, collapsed, but then stood up. As the bull moved downhill again with wolves attached to his rear end and hind legs, breeding male 21 leapt up and grabbed the right side of his neck.

Wolf 21 stood on his hind legs trying to hold onto the neck but lost his grip. The bull headed downhill, and the wolves followed. As he tried to run downhill, the bull fell again near remains of an older bull elk. The wolves were on him right away, with 21 biting at his throat. The bull arose and thrashed around with his antlers, but the wolves easily avoided them. He collapsed but arose again. Wolf 21 tried to grab the elk's throat again. The wolf reared up on hind legs and put a front paw on the bull's shoulder as he tried to bite his neck. A black came over and leapt up at the throat. Now both 21 and that black bit the neck, standing on their hind legs. Other wolves were biting at the rear end at 0904 hr.

The bull continued to thrash around with his antlers, but it did him no good. Some wolves bit his rear end, and a black bit his side. The bull turned and lunged at the wolves with his antlers. Wolf 21 and the black lost their grip on his neck. At least nine wolves stood nearby

not helping. The bull stood up with several wolves next to him, none at his throat. Then a black casually walked around, jumped up and bit his throat but could not hold on. Another wolf was on a hind leg, and a third bit his side. The bull slowly continued downhill, then collapsed again. Wolf 21 was behind him but did nothing. Wolf 42 was also there but not attacking. The bull arose and continued to move slowly downhill. Several wolves bit his hind legs, rear end, and sides. At 0905 hr, 21 bit the bull's throat. He maintained his grip this time, and the bull collapsed. Wolf 21 adjusted his hold on the neck, and the bull raised his head and struggled to arise.

Wolf 21 continued holding the elk's throat. Most of the other wolves did nothing. The bull put his head down at 0907 hr and appeared dead. That was about 4 min and 23 sec after the bull first fell and the wolves started their attack.

8. January 26, 2003—Yellowstone Park, Wyoming (R. McIntyre)

At 1043 hr, breeding male 21 led seven Druid wolves west and passed between two small bunched elk herds north and south of his route. Breeding female 42 was bedded east of him. Wolf 21 stopped and looked back at the other wolves, then at the elk north of him. He continued west. Suddenly, 21 ran downhill northeast toward the nearest elk herd, the black pup with him. The pup soon passed 21, and they both continued toward the elk. The herd ran east, then fanned out. A black wolf and the gray pup ran at the elk from the south. A subgroup of elk went west, and the gray pup chased it. Wolf 42 chased another subgroup going southeast. Wolf 21 slowed down and was way behind the elk and other wolves. A black wolf and the gray pup were running south after an elk subgroup. They targeted a calf near the rear of the group. Wolf 42 came in from behind those two wolves. The calf tried to keep up with the other elk but fell behind.

The herd ran southeast, then fanned out. The calf slowed. The black male grabbed the calf's rear and pulled it down. The gray wolf and the gray pup both bit the calf's throat. Wolf 42 attacked its hindquarters. A black wolf ran in from the north and the black pup from the west. All five pulled the calf down. Wolf 21 slowly arrived from the south at 1049 hr. A pup ran in but hesitated for a moment. The calf jumped up and ran off, but the black male easily grabbed it again. The pup rushed in, leapt up,

FIGURE 4.18. Even when cornered with no place to go, elk can rebuff wolves, as these surviving elk attest. Agate Creek Pack, 2005, Yellowstone National Park. (Photo by Daniel Stahler.)

and grabbed the throat. Wolf 42 came in. The black male and the gray pup both bit the throat while 42 attacked the hindquarters. Another black wolf, most likely 253, and a black pup ran in to join the group. Together, all five wolves attacked the calf and pulled it down. Wolf 21 slowly made his way to the group of five, and all ate side by side (1049 hr).

9. March 2, 2005—Yellowstone Park, Wyoming (D. Stahler)

During aerial tracking of the Agate Creek Pack (six wolves: 113M, 472F, 383M, 471F, and two noncollared blacks), I located the pack on top of Bumpus Butte. I spotted an adult cow and two calf elk standing on a cliff ledge of the butte overlooking the canyon, with the Yellowstone River many meters below. I seemed to have come on the end of the encounter, as most of the pack was leaving the elk. Two younger wolves, however, were approaching the cow and calves within about 3 m. The cow was between the two calves and the wolves, stomping her front legs while standing her ground. The calves were looking down over the cliff as if assessing an escape route (fig. 4.18). We circled for about 10 min as the two noncollared black wolves walked in and out of the conifers, looking to angle in from both sides of the elk. The cow stood calmly, with an occasional stomp and ears-forward posture at the approaching wolves. The wolves appeared hesitant to engage the cow further and eventually walked away to catch up with the rest of the pack, which had traveled off.

10. March 6, 2005—Yellowstone Park, Wyoming (M. Metz and D. Smith)

At 0646 hr, we observed 25 wolves from the Leopold Pack traveling ~2 km from their previous bull-elk kill. After about 10 min, the wolves disappeared for ~40 min before we briefly observed two wolves chasing a single bull elk at 0736 hr.

FIGURE 4.19. Adult male Wolf 287 (*top*) and a pup of Yellowstone's Leopold Pack in 2005 pursue a bull elk that had recently lost his antlers across Blacktail Deer Plateau. During the early part of this interaction, the wolves attacked a group of bulls and quickly targeted this antlerless elk even though he appeared to be more vigorous than the others. (Photo by D. W. Smith.)

We then watched the wolves travel for ~15 min before they headed into a draw and flushed ~15 bull elk. Three wolves chased a subgroup of two of these bulls, one of which still had its antlers. The two bulls stood their ground and grouped while three wolves circled them. The anterless bull began turning in place and kicked at the wolves, as they now seemed solely focused on it. After about 1.5 min, a fourth wolf joined in as the two elk continued to stand their ground. Meanwhile the wolves repeatedly darted in and out of the vicinity of the elk to get them running, but the anterless bull was the only one that charged and kicked at the wolves.

After another 30 sec, there were seven wolves. The anterless bull continued to charge the wolves and turn in place, and the wolves were unable to get at his hind end. This continued for another 2 min (total duration now 4 min). More wolves continued to arrive, and the anterless bull fled (fig. 4.19). One wolf grabbed the elk's hind end, and the elk kicked and turned to attempt to antler the wolves (although no antlers!).

The elk continued running with eight wolves now right behind. Little snow covered the ground, and the elk disappeared in a draw with seven wolves in pursuit. By then the aircraft crew could follow the chase. Only two wolves pursued, adult male 287 and a pup. The adult male contacted the hind end of the elk, but the elk continued to run. Visibly tired, 287 and the pup then gave up after another minute. The interaction lasted 13 min, and the chase covered almost 3 km over mostly snow-free ground.

11. March 15, 2005—Yellowstone Park, Wyoming (M. Metz)

At 0845 hr, the 25 Leopold wolves began to travel and almost immediately encountered four bull elk that grouped and stood their ground. Some wolves watched the elk, and one briefly approached the bulls, but the pack mostly just traveled by and disappeared at 0900 hr.

At 1011 hr, four wolves, including breeding male 534 and breeding female 209, arrived near the traditional Leopold den. At 1016 hr, these four bedded. The rest of the pack was more than 500 m behind the four when 29 bull elk began running west toward the den in response. Four other bull elk did not run and remained in the forest. The four wolves arose, and at least two wolves from the lagging group began chasing the bulls but quickly gave up.

The elk continued to move toward the four wolves, which moved out of the way of the elk. The elk then stood their ground. As this interaction was concluding, one wolf began chasing four other bulls upslope through trees. The elk grouped and briefly stood their ground, but then an antlerless bull fled, followed by two of the three antlered bulls. Two black wolves and a gray chased these three elk. One antlered bull split away, and two wolves targeted the antlerless bull. That bull then ran downslope toward most of the wolves, and three wolves chased him through the timber. The elk began stotting, and more wolves joined the interaction. The elk stood his ground and charged the wolves, which were darting in and out at the elk. The elk continued to stand his ground and kick at them. The wolves got the elk running, although he again quickly stood his ground, with 10 wolves around him. He charged and attempted to antler the wolves (although no antlers). As an 11th wolf joined in, the elk ran, and some wolves contacted his hind end. The elk kicked his rear leg and again tried to butt the wolves away. He then fled with the wolves in pursuit, kicking at the wolves as some grabbed his hind end. He stopped, turned in place, kicked one of the wolves in its side and stomped another before again running.

At least 16 wolves were chasing the elk, which kicked his rear legs at times while running. At one point, one wolf briefly grabbed the elk's rear leg. The elk appeared to limp slightly as he ran. The elk then joined a group of at least eight bulls (from the group of 29), which stood their ground, and the wolves quickly give up pursuit of the bull. Attempt lasted about 4 min.

12. May 27, 2005—Yellowstone Park, Wyoming (M. Metz)

At 1008 hr, we observed four wolves running into a group of at least 21 elk, which included one neonate calf. The mother immediately chased off three of the wolves. After being charged, two of these wolves again quickly ran toward the calf and grabbed it by the neck and hind end, injuring it before letting go as the mother again charged them. While these two wolves were attacking the calf, the mother had been chasing another of the wolves, giving these two a chance to grab the calf. After the two wolves were chased off, three darted back in, grabbed the calf, and went out of sight.

Although the wolves killed the calf at 1009 hr, we observed the mother until at least 1243 hr. During this period, she paced around and looked to the area out of our sight where the wolves were feeding. She also went to where the wolves first grabbed the calf. At 1240 hr, one wolf came out from the kill site, and the cow immediately chased it for 1 min before they both disappeared.

13. August 3, 2006—Yellowstone Park, Wyoming (D. Smith)

While aerially locating three radioed adult male and one adult female wolves, we found four Mollie's Pack wolves chasing an antlered bull elk in Pelican Creek Valley. Viewing conditions were difficult. The elk and wolves were in the creek, and the bull was running up the creek with wolves chasing. Instead of continuing up the creek to an opener grassland, the elk turned and ran down the creek with the wolves following closely. Several times the bull faced the wolves, and they jumped at him biting. He fled but the wolves bit him and hung on for several seconds. The female was not attacking as vigorously. The wolves got the elk down on his side in the water, and we thought that it was over, but the elk struggled back to his feet and continued running again—still not leaving the water but wounded. All four wolves inflicted more bites and hung on for longer. The wolves pulled the bull down in the creek, holding on and biting, and it succumbed (figs. 4.10–4.13).

Later, we found about a dozen elk skulls of different freshness (most not recent) and mostly antlered, suggesting an area of frequent summer kills, a terrain trap?

14. March 22, 2007—Yellowstone Park, Wyoming (D. Smith, C. Benell, N. Legere, L. Williamson)

At 0812 hr, we aerially spotted the Leopold wolves attacking a lone, antlerless bull elk. A yearling male and five probable adults were involved, but moments later nine wolves were attacking as the bull fled downhill.

FIGURE 4.20. A yearling Yellowstone wolf chased this bull elk, which had recently lost his antlers, for over 3 km even though the rest of the pack had quickly given up. The same wolf chased another elk a similar distance the next day. Both hunts were unsuccessful. Yearling wolves have high participation rates in hunts, possibly because they are honing hunting skills. (Photo by D. W. Smith.)

Thirteen wolves then followed, which included the breeding pair plus several other adults and some pups. By 0814 hr, only the black yearling male and a gray were left chasing the bull, running full speed. Shortly after, the gray quit chasing (0815 hr), leaving only the black in close pursuit (fig. 4.20). The chase continued for a straight-line distance of about 3.2 km, although the route was not straight. Terrain was rolling hills, mostly snow free, and predominantly downhill. The wolf kept his tail high, appearing to use it for balance as he ran over uneven terrain. At one point, he crossed a hardened snow field, fell, and slid to the bottom, where he quickly regained his feet and the chase. The elk looked fatigued, as its mouth was gaping open. The wolf eventually caught up and ran alongside the elk without attempting to attack. The wolf looked over frequently while the elk did not and kept running. Without warning the wolf stopped chasing (0817 hr), and the elk continued.

15. December 9, 2007—Yellowstone Park, Wyoming (D. Smith, K. Cassidy, T. Wade)

While aerially locating the Druid Peak pack (1009 hr), we saw a bull elk running through the trees pursued by several wolves. (The ground crew saw the beginning of the chase with three blacks and one gray in pursuit.) Quickly the chase advanced with six females—four yearlings and two adults—pursuing (fig. 4.4). Neither adult male (302 or 480) was involved. The chase continued for about 3.2 km over level terrain, 30 cm deep, fluffy snow, and in and out of forest with some sagebrush. The bull encountered a steep embankment, stopped, and faced the wolves; by this time only an adult and yearling female were left. The bull was tired, and his breath was pulsing out at frequent intervals. At this point both females left the bull to wander off (1016 hr). At one point, it appeared that the wolves bit the elk, but we could not be sure, and he looked uninjured.

PLATE 1. This moose stood its ground for 5 min, then the wolves gave up (account 15; Mech 1966a).

PLATE 2. Wolves tend to attack primarily moose that run from them, although they do not attack all those that run. (Photo by R. O. Peterson.)

PLATE 3. Usually wolves tend to bite the rump of a moose first and most tenaciously. (Photo by R. O. Peterson.)

PLATE 4. The Leopold Pack chases a bull elk across Blacktail Deer Plateau in Yellowstone National Park, 2007. This is typical behavior for elk when approached by wolves, and wolves prefer to attack fleeing elk because standing elk are more dangerous. (Photo by Matthew Metz.)

PLATE 5. A much more difficult situation for wolf attack. Standing prey are better able to defend themselves, lashing out with front legs and even antlers; wolves are wary and reluctant to attack such prey. Leopold Pack in Yellowstone National Park, 2007. (Photo by Matthew Metz.)

PLATE 6. Six Druid Peak Pack females chasing a bull elk in Lamar Valley, Yellowstone National Park, 2007, assessing capture opportunities. The two large males (302 and 480) in this pack did not participate in the hunt for unknown reasons, but the other wolves pursued the elk until he stood his ground, where the female wolves immediately stopped and retreated without further attack. Usually, this would have been the point where the larger males might have pressed the attack and gotten the tired elk running again (the females had chased it for approximately 3 km) leading to a kill. In this case, the elk went unscathed. (Photos by D. W. Smith.)

PLATE 7. Severe winters with deep snow constitute a major vulnerability for elk and other ungulates, usually making the prey easier for wolves to kill. Slough Creek Pack, Yellowstone National Park, 2008. (Photo by Daniel Stahler.)

PLATE 8. Bechler Pack wolves in 2007 attempting to capture an elk from a group of two cows and a calf that have retreated to water. Refuge in water is a common strategy for elk in Yellowstone National Park. A disproportionate numbers of kills are found near water. This hunt was unsuccessful. (Photo by Matthew Metz.)

PLATE 9. Elk retreat to water upon wolf attack even when the water seems too shallow to offer an advantage, as this bull elk in Pelican Creek, Yellowstone National Park in 2006 demonstrates. (Photo by D. W. Smith.)

PLATE 10. This bull elk ran back and forth in the same stretch of stream while the wolves attacked continually, eventually killing it. Attack points were at the rear. (Photo by D. W. Smith.)

PLATE 11. Another example of when running to water is a poor strategy. Why elk flee to shallow or partly frozen water is unknown. Three-yr-old (prime-aged), dominant male Wolf 295 of the Swan Lake Pack (lighter wolf) led this attack with help from another prime-aged male. The 7-yr-old, dominant, female Wolf 152 stood on the shoreline and watched the attack. Yellowstone National Park, 2005. (Photo by D. W. Smith.)

PLATE 12. Post 1988 forest-fire evidence suggests that elk preferentially seek out burned forest, possibly because the downed trees present more difficulty to the pursuing wolves than to the fleeing elk. Leopold Pack, Yellowstone National Park, 2007. (Photo by Matthew Metz.)

PLATE 13. Even when cornered with no place to go, elk can rebuff wolves, as these surviving elk attest. Agate Creek Pack, 2005, Yellowstone National Park. (Photo by Daniel Stahler.)

PLATE 14. Adult male Wolf 287 (*top*) and a pup of Yellowstone's Leopold Pack in 2005 pursue a bull elk that had recently lost his antlers across Blacktail Deer Plateau. During the early part of this interaction, the wolves attacked a group of bulls and quickly targeted this antlerless elk even though he appeared to be more vigorous than the others. (Photo by D. W. Smith.)

PLATE 15. Deep snow definitely gives wolves an advantage. (Photo by Daniel Stahler.)

PLATE 16. Elk bog down in deep snow, so once wolves catch up to them, killing can be easier. (Photo by Daniel Stahler.)

PLATE 17. Sitting and waiting is probably more common for wolves than presumed, especially with less vulnerable prey, such as a bull elk. When killing quickly is too dangerous, wolves may attack, wound, back off, and wait for the prey to weaken, as has long been known for moose and bison. It took wolves several days to kill this elk in Gardner's Hole of Yellowstone in early winter 2011. In this case, the dominant female wolf of the Canyon Pack slept unconcernedoff to the left. (Photo by D. W. Smith.)

PLATE 18. Dall sheep often escape wolves by taking refuge on pinnacles or precipices that wolves can't reach. (Photo by T. J. Meier [Mech et al. 1998].)

PLATE 19. The interaction between wolves and bison is the last living example of the megafaunal predator-prey interactions that typified the North American plains during the Pleistocene epoch more than 10,000 years ago. (Photo by Daniel Stahler.)

PLATE 20. Wolves attacking a mature male bison in Yellowstone National Park. (Photos by D. W. Smith.)

PLATE 21. Bison calves move to the center of the herd in response to wolves. (Photos by Daniel Stahler.)

PLATE 22. Bison herd on large geothermal area watching arrival of Mollie's Pack in Yellowstone National Park, March 17, 1999. Accounts 3, 6, and 7 refer to this area. In winter, wolves are often reluctant to attack large herds on open patches, preferring to attack them as they move single file through deep snow between patches. (Photo by Kevin Honness.)

PLATE 23. The conditional nature of bison mutual defense. Bison commonly defend one another, but only when the risk of predation to themselves is low. In the example above, snow around where wolves have captured a bison deters others from helping it due to the difficulty of defending against wolves in snow. (Photo by D. W. Smith.)

PLATE 24. When confronted by wolves, musk oxen face them. (Photo by L. D. Mech.)

PLATE 25. As wolves persist, musk oxen tighten their group. (Photo by L. D. Mech.)

PLATE 26. After considerable harassment by wolves, musk oxen may flee (account 20). (Photo by L. D. Mech.)

PLATE 27. Calves are easier for wolves to catch and kill if the wolves can separate than from their mother or the rest of the herd. (Photo by L. D. Mech.)

PLATE 28. Arctic hares can run about as fast as wolves but can zigzag more sharply. (Photo by L. D. Mech.)

16. December 27, 2007—Yellowstone National Park, Wyoming (R. McIntyre)

0722 hr—16 Druid wolves were bedded on a high bank. They arose and howled at 0730 hr. Some looked down the creek to the west where there were several bull elk.

0740 hr—One bull climbed up the high bank and continued south across the middle flats. The wolves looked at him. At 0743 hr, seven wolves got up and ran after the bull. The group of seven included the breeding female 569, two blacks, and four grays. The bull stumbled, but continued south at a good pace. He moved more quickly than the wolves.

Four blacks—including second-ranking male 302 and dominant male 480—and a gray remained bedded on the bank. The chasing wolves gave up and returned to the south bank. Several wolves looked down at the other bulls, then ran down toward them. Four antlered bulls grouped and stood their ground.

0746 hr—One of the black males (either 302 or 480) on the bank sat up and ran downhill. The other followed. One bull separated from the other bulls, and the wolves chased him. He climbed the high bank and went east. The wolves easily kept up and clustered at his hind legs as he ran.

0747 hr—A black grabbed at a hind leg and was kicked off. The bull ran downhill to the flats. Three grays and three blacks led the chase. One black fell, arose, and continued running. A gray tried to grab a hind leg and was almost kicked by the other hind leg. Others were attacking the hind legs as the bull kicked back at them. A black held on to a hind leg, which began to slow the bull. Wolves bit at his sides. Wolf 302 jumped up and got a holding bite on the bull's upper throat. The wolves pulled him down. He got right up. Wolf 302 continued to hold the throat, while 571 leapt up and grabbed the other side of the throat. The two of them pulled the bull down by the throat. Other wolves attacked the hindquarters and sides.

0750 hr—Wolf 569 bit the side of the neck, then let go. Wolf 571 lost her hold but jumped up again and bit where 569 had bitten the bull. The wolves pulled the bull down, but he got up. He went down again but arose. He tried to lunge at the wolves with his antlers, went down, and could not get back up.

0751 hr—Two wolves were on the bull's neck. He struggled to get up but could not stand. At 0753 hr, his head was still up. He almost arose but failed. All the wolves bit at him. The bull tried to arise at 0755 hr but collapsed (www.press.uchicago.edu/sites/wolves, video 12). Wolves were feeding at that time.

17. February 6, 2008—Yellowstone Park, Wyoming (R. McIntyre)

The Druid Peak wolves were going west at 0858 hr. Elk were scattered throughout that area. Some wolves ran at the elk several times. Around 0932 hr, two adult female wolves ran at a cow and calf, which fled east along the slope. The two elk swung downhill to the south, cow ahead. One wolf led the chase. At 0933 hr, the calf stumbled in deep snow, and that wolf grabbed the calf by the back of the neck before it could arise. The other wolf ran in and bit the hindquarters. The calf arose, and the first wolf lost her grip. She bit the side of the neck, then let go. Both wolves attacked the hindquarters as the calf struggled to break free and run downhill.

Male 302 ran in directly to the front of the calf, jumped up, and grabbed it by the upper throat. The wolves pulled the calf down. A gray pup ran in. At 0935, one wolf sidestepped away. The calf stopped struggling (www.press .uchicago.edu/sites/wolves, video 13).

18. February 18, 2008—Yellowstone Park, Wyoming (D. Stahler)

I aerially located 12 members of the Slough Creek Pack surrounding a lone cow elk in timber. She was standing her ground against about five wolves. As the others approached, the cow fled, plunging into deep snow in an open meadow. One wolf quickly closed in directly behind her. Although the snow was light and fluffy, its depth slowed the cow considerably. After about 50 m, three wolves came within reach of contact. The lead wolf grabbed her hind leg above the hock slowing her as two more wolves grabbed on. The cow dragged the wolves through the snow but was slowing down with two wolves on hind legs and a third attacking her right side behind the front shoulder. The rest of the pack caught up, and with two wolves at the neck and five at the back end, the cow dropped into the snow. Several pups (10 mo old), were part of the takedown. The deep, soft snow appeared to offer a relatively safe hunting situation to avoid potentially dangerous kicking by the cow for less-experienced hunters (figs. 4.21–4.23).

FIGURES 4.21–4.23. Deep snow definitely gives wolves an advantage. (Photos by Daniel Stahler.)

19. May 24, 2008—Yellowstone Park, Wyoming (K. Cassidy)

At 0940 hr, the Slough Creek Pack of eight wolves was near a cow elk. They chased her for about 40 m, but she stayed near patches of upland grass and charged the wolves for 20+ m at least eight times. The wolves fanned out around the cow. Several seemed more interested in the grassy areas, sniffing around and not paying attention to the cow. At 0941 hr, male 30 sniffed around and came to within a meter of a calf. The calf jumped up, and the wolf was on it within one step. All eight wolves converged on the calf and began feeding. The cow continued to charge the wolves, and several chased her about 50 m as the others fed.

20. May 31, 2008—Yellowstone Park (K. Cassidy)

At 0738 hr, seven Druid wolves reached thick conifer regeneration and spread out, traveling slowly and sniffing around. At 0750 hr, the breeding male reached a small elk calf on the ground. He nudged it with his muzzle, and when it moved he grabbed it by the back of the neck. The cow came out of the trees and chased him. As the wolf tried to get the calf over a downed log, he dropped it. A 2-yr-old female wolf directly behind him grabbed it by the throat and got it over the log as the cow charged and stomped over both the wolf and calf. The male grabbed the calf again by the middle and ran into thick conifers. The remaining wolves and the cow followed at a run. The wolves with calf stopped about 20 m in the forest, and the wolves seemed to be feeding on the calf. At 0752 hr, the cow elk got close to the wolves, and two gray yearlings chased her about 300 m south. The cow returned immediately and charged the wolves five times. By 0827 hr, wolves were slowly traveling uphill, at least one carrying the hindquarters of the calf.

21. March 3, 2009–Yellowstone Park, Wyoming (K. Cassidy)

At 0721 hr, all 13 Druid Peak wolves were traveling east, in and out of sight until reaching a long, open ridge at

0830 hr. At 0831 hr, they spread out and ran upslope along a windswept ridge toward 15 bull elk (only some antlered). The bulls ran upslope for about 100 m when the last bull (anterless) fled to the north into deeper snow.

The three lead wolves (all 2-yr-old females) followed the bull, sometimes bounding through the snow alongside him. The other bulls stopped running almost immediately and spread out along the ridge, watching the chase. The targeted bull reached a windswept ridge but slowly ran through it into deep snow again. Three steps into the snow, and one lead wolf grabbed the bull by the low neck/shoulder area at 0832 hr. The bull turned back around, jumping through the deep snow and headed back toward the way he came. The three lead wolves were now joined by another 2-yr-old female, a yearling female, and two male pups. The two pups stayed farthest back as the bull and five females ran quickly downhill.

The breeding male cut downhill into the path of the bull, but the bull ran past, and all continued running, with the male and a 2-yr-old female ahead. An additional black pup tried to cut downhill like the breeding male but was too late and joined the chase about halfway back. The bull continued downhill, alternating running through windswept areas and deep snow and separating itself from the wolves. Three more wolves (not part of the original contact—a yearling female, a male pup, and a female pup) cut downhill right behind the bull. The breeding female was about 30 m to the right and must have cut downhill farther back than the rest. The bull hit some deep snow but headed directly toward a small windswept area with the cornice to the north. The lead wolves, mostly pups, stopped several meters away as the breeding male caught up and stepped in front of them toward the bull's head. The breeding female moved around to the downhill side of the bull as 10 of the wolves made a semicircle around him and made short lunges toward him.

At 0833 hr, the bull faced the wolves. Behind him the ridge dropped off, creating a cornice. The bull stomped his front feet several times and kicked the breeding male on the back with both front feet, sending him downhill several steps. Immediately after, a black grabbed the hind end of the bull. The bull spun around, but a gray grabbed his neck and many others latched onto his rump. The bull fell but slid into deep snow and then off the cornice at 0834 hr. The wolves let go and watched the bull slide down for a second, and then several jumped off the 3-m

cornice as others found a less steep way. The bull stopped sliding after about 50 m as he hit several trees. He arose and ran downhill just as the lead wolves caught up. He ran about 50 m downhill through deep snow with the wolves behind and alongside. Several grabbed the bull, and he fell onto his side in the deep snow and started sliding downhill. He picked up speed, sliding on his back with the wolves chasing behind. He had his head up and was flailing around but continued to slide at least 40 m before hitting a stump and stopping abruptly at 0834 hr. The wolves were on him immediately as he slid another 3 m onto a small flat area.

At 0838 hr, the bull struggled and then jumped up. The wolves all grabbed him, and he dropped again. I think he was dead by 0840 hr as most of the wolves started feeding (www.press.uchicago.edu/sites/wolves, video 14).

22. October 14, 2009—Yellowstone Park, Wyoming (K. Cassidy)

At 0850 hr, nine Druid Peak wolves (six with class 3 mange) milled around while a tenth, a black 3-yr-old female, was halfway down a bench toward a river stalking something. She started to run, and an elk jumped into the river. At 0854 hr, the wolf swam toward the elk in the middle. The wolf turned around and got out, stood on the bank, and watched the elk in the water. The other nine wolves came running in. Two other 3-yr-old females jumped in and chased the elk. The elk continually went up and downstream, going from about 30 to 90 cm of water. It spun as the wolves got close, and they all splashed up lots of water. Four 3-yr-old females and three yearlings were in the water. A 2-yr-old female, yearling male, and a pup stood on the banks and watched. The wolves in the water made 32 unsuccessful attempts to grab the elk by the rump, flanks, or throat. Several emerged from the river and stood on the bank shivering before jumping in again. At 0920 hr, most of the wolves were out on both banks trying to dry off by rolling in brush or licking each other. The elk stayed in the river.

The original 3-yr-old female began another attempt at 0940 hr. She was joined by two others for four capture attempts. The wolves again got out and rested. Many of the wolves were still standing and shaking. By 1045 hr, the wolves headed away from the river to the south. The elk remained in the water until I departed at 1108 hr.

23. December 12, 2009—Yellowstone Park, Wyoming (K. Cassidy)

About 0900 hr, the Lava Creek Pack of three wolves were traveling east through sage hills and slopes. A snowstorm moved in, and I lost visibility. At 0902 hr, the snow lifted, and I saw the gray 3-yr-old female chasing 22 cows and calves. She seemed locked, in chasing a specific cow as a group of 15 split off. The targeted cow split from a group of seven. The two larger groups stopped running as the single elk continued with the wolf immediately behind. The elk kicked back with her hind left leg (0903 hr), but the wolf dodged it. The cow tripped about 40 m later, and the wolf tried to grab the rear right leg, but the cow arose quickly and continued running. About 3 m separated the two as they fled toward the river. Both disappeared, but the breeding male ran into the same area at 0904 hr.

We moved to see better and, at 0907 hr, saw the elk go down very close to where we lost visibility earlier. The original wolf and the breeding male bit the elk's flanks, and the white breeding female held her throat with the elk's head on the ground. The elk stopped moving. At 0910 hr, the two wolves on the sides were feeding, and the white female released her throat.

24. March 25, 2011—Yellowstone Park, Wyoming (C. Anton)

At 0742 hr, we watched three adults and four pups of the Lamar Canyon Pack traveling, and pup 776F headed toward a lone, cow elk. As the wolf approached to within 50 m, the elk charged and foot stamped at her. After this happened several times, it seemed like the wolf was trying to get the cow into the deep snow and away from bare ground. However, time after time the elk charged just to the edge of the snow and quickly retreated to bare ground. Several minutes later, a second pup joined, and both tried to get the cow to chase them into the snow. This approach was unsuccessful, and after five or six attempts, the wolves gave up.

25. May 2012—Jasper Park, Alberta, Canada (D. Dekker)

On a semi-open island in the braided Snake Indian River, a herd of grazing elk suddenly became alert. Ears cocked, all faced in the same direction. The first to flee were four cows and two calves. Seconds later, the main herd fol-

lowed, plunging through the water and heading for the wooded shore. Three black wolves streaked across the island and splashed through the shallows to overtake the last of the herd. One cow dodged and stood in the main channel. When a wolf lunged at her rump, she kicked with her hind legs and tried to evade her attacker by crossing and recrossing the river. She halted where the water was belly deep and turbulent. On the bank, the three wolves paced until one entered the water and swam toward the elk, but she waded out of reach across the strong current. Unable to get close to his prey, the wolf found some footing on a shallow spot or submerged boulder. Only his black head remained visible above the choppy waves. He kept up the siege for some 15 min, then rejoined his pack mates and wandered off. The cow stayed in the river for a long while, until she waded back, stiffly and hesitantly, to the island.

26. July 7, 2012—Yellowstone Park, Wyoming (R. McIntyre)

About 0930 hr, I saw a spike bull in the river up to a bit above his belly. A gray wolf was on the north side near the elk. Female yearling 820 was east of that gray. South of the elk were Lamar breeding female 832, male 754, breeding male 755, and another gray. Most were bedded. Later, four more wolves found Wolves 832 and 754 and the elk.

At 1046 hr, 832 entered the water and swam upriver toward the elk, and 755 did also. The bull stood his ground. The current was fast there, and it hindered the two wolves as they swam at the elk. Wolf 832 emerged on the north side and got closer to the bull. She bounded through the water at the bull. Both wolves then swam toward the bull against the current. Wolf 755 came out, then went back in, while 754 and a gray stood and watched from the bank. Wolves 832 and 755 tried to fight the current as they swam upriver toward the elk. The bull bounded past them faster than they could swim. The wolves swam after him as he waded downriver.

The bull headed back upriver as the wolves approached and easily passed the two. The wolves repeatedly tried to swim after the bull, but he continued to rush past them as he waded either upriver or downriver through the deep water. Wolf 754 and one of the gray adults briefly entered the water, attempting to attack the bull, then emerged and rested. The breeding pair continued to fight the current to get at the bull but eventually gave up and came out. The breeding pair made repeated attempts at the bull for 28 min, from 1046 hr to 1114 hr.

Wolf 832 went back into the river at 1202 hr and bounded through the water downriver at the bull. He waded downriver faster than she could swim. She emerged and went right back in the river and swam at the elk. The bull went upriver again. Wolf 832 came and went back in, bounding through the water after him, and 755 ran into the river. The bull turned around and ran past the breeding pair. Wolf 754 and the other three grays headed parallel to the elk downriver on the bank. The pair swam after the bull. When 832 left the water, 755 continued to swim upriver after the bull and then bounded through the water after him. The bull turned and tried to bound past 755, but 755 managed to bite his hindquarters and hang on as the bull tried to escape downriver. The wolf lost his grip. Wolf 832 entered the river and bit at the left side of the bull, and 755 bit his shoulder. Wolf 832 let go and grabbed the bull's neck. They pulled him down into the water, and 832 let go of her bite, but 754 ran into the water, swam over, and bit the bull's throat. The other three grays entered the river and helped in the attack. The six wolves had to swim as they bit the elk. One gray was biting at his back.

The elk's head dropped into the water (1207 hr). A younger gray swam off downriver and another swam off upriver. A third, younger, gray swam off. Wolf 820 was one of those grays, and she came out of the river. That left the breeding pair and 754, the three oldest wolves, to continue the attack. They held the bull down in the water. A younger gray returned and helped the other wolves. All four pulled the carcass toward shore as they swam around the elk, which now seemed dead (1208 hr).

Conclusion

The above accounts of wolves hunting elk are a selection of observations by 13 observers, mostly from Yellowstone but also from Jasper National Park. All of the observations from Yellowstone have information available on pack size and environmental conditions, two key factors in interpreting hunting success and killing strategy. For example, the winters of 1996–97 and 2011–12 were exceptionally severe, and the winter of 2007–8 was

of average severity, while all of the others between 1995 and 2012 were below average, and this variability affected hunting behavior.

Overall, and after reading all of the hunting accounts provided to the authors (most not included), we conclude that wolf strategies while hunting elk are highly variable, dependent on conditions (most particularly snow and availability of water), season, pack composition (pups, adults, and male vs. female), and what seems like spontaneous decision making based on events at the moment. This last point is especially relevant because, as discussed in chapter 3, many have hypothesized that wolves use various strategies when hunting (Muro et al. 2011), and Mech surveyed 19 wolf biologists about this question (Peterson and Ciucci 2003). We were struck by the number of accounts where many wolves appear to cooperate in a hunt, but quick decisions based on what other pack members do, or elk do, seem to be the norm, suggesting no planning of hunts. And sometimes wolves split into separate hunting groups, with each group unaware of what the other is doing; they may even make two kills in the process (account 3). It is apparent that each of the many accounts of wolves hunting elk was in some way unique, which may prevent strategic cooperation. As mentioned in chapter 3, to detect a strategy would take an analysis of a large number of similar hunts, yet hunts are usually dissimilar.

Nevertheless, some generalizations of wolf hunts of elk can be made. When elk flee, it tends to be downhill rather than up. Four is the optimal number of wolves actually involved in an elk attack, even though far more pack members often accompany them (MacNulty et al. 2012). Although the number of wolves observed hunting elk was one to 20, any more than four did not improve hunting success. Further, wolf age, size, and gender (males being heavier) are also important, so pack composition makes a difference during wolf-elk interactions. Wolves mature rapidly and reach peak physical ability to catch elk when the predators are 2–3 yr of age (MacNulty et al. 2009a). After that, ability to catch elk declines, although wolves do continue to hunt successfully for the rest of their lives. Larger packs allow older wolves to participate selectively in hunts. When full grown, male wolves are about 20% heavier than females (Butler et al. 2006; Mech 2006)—a difference

FIGURE 4.24. Sitting and waiting is probably more common for wolves than presumed, especially with less vulnerable prey, such as a bull elk. When killing quickly is too dangerous, wolves may attack, wound, back off, and wait for the prey to weaken, as has long been known for moose and bison. It took wolves several days to kill this elk in Gardner's Hole of Yellowstone in early winter 2011. In this case, the dominant female wolf of the Canyon Pack slept unconcerned nearby off to the left (Photo by D. W. Smith).

that also affects hunting ability, as noted above. Young wolves being lighter weight tend to be faster and better at pursuing prey than larger (older) males (figs. 4.3, 4.5, and 4.6). Thus, these speedier wolves tend to attack vulnerable prey while the larger but slower adult males that catch up tend to more effectively bring the prey to the ground (MacNulty et al. 2009b). Indeed packs with one large male had greater hunting efficiency than packs with none, although those with more than one large male did no better (MacNulty 2007). Pups more often tend to hunt when prey are more vulnerable (during severe winters or late winter) or when pack size necessitates it (account 4), but they also may participate during any hunt.

The above generalizations do not mean that wolves always hunt this way or that wolves without the optimal sex/age balance cannot kill prey. Wolves are exceptional at adapting (fig. 4.24).

In general, hunts are short, in that wolves decide quickly whether an elk is worth pursuing, and as a result the chase is not long, but there is wide variation. Carbyn's (1974) average chase distance was 351 m, and in Yellowstone it was 978 ± 142 m (Kauffman et al. 2007). Most hunts end quickly, but some last for hours (typically those of bulls; fig. 4.22). Calves were preferred. The initial attack points are anywhere a wolf can grab, which is commonly a hind leg or fleshy area on the side, but very soon after the initial bite, wolves grab the neck, especially on calves and cows. In forested areas observers noted that vegetation was used to screen wolves away from the backside of elk. We did not see this in Yellowstone, probably because of the openness, but we did find evidence of this strategy at kill sites. In late winter, when bull elk have dropped their antlers, it appears wolves select them even though they may be in better condition than antlered elk (YNP unpublished data; accounts 10 and 11).

Certain terrain features also favor wolf attack. Kauffmann et al. (2007) found a disproportionate number of elk killed near streams. In one such stream area in Yellowstone, Smith aerially observed wolves making a kill and, when retrieving the remains, found about a dozen kills, mostly not recent (account 13). This suggests a location favorable to wolves for killing elk, or as we have come to call them, a "terrain trap." Landis (1998) has found many similar areas (see film sequences). These observations support the hypothesis that weaker prey tend to flee to water and downhill (Bibikov 1982).

5

Mountain Sheep and Goats

OF ALL THE wolf's hoofed prey, perhaps the mountain sheep and goats challenge the wolf's physical agility more than any other. Both prey are specialized to inhabit steep mountains with rocky promontories, craggy peaks, and jagged boulder fields. Maneuvering in this foreboding terrain is second nature to these creatures, and they are physically well adapted to it. Wolves, however, are more generally adapted and cover a wide variety of terrain, from swampy areas and brushy forests through tundra and taiga to mountain tops, where they prey on an extensive array of species, each with its own specializations. Little is known about the interactions between wolves and mountain goats, especially those of Eurasia. However, we can assume that what is known about their interactions with sheep and goats in North America pretty much applies everywhere. Adolph Murie (1944) conducted the most extensive study ever of wolf-sheep interactions in Denali National Park (then Mount McKinley National Park) and provided lasting insights into that relationship, which since have been confirmed by various pieces of anecdotal information from Denali and other areas.

Mountain Sheep

Although several species of mountain sheep inhabit wolf range, the basic body form and behavior of sheep relevant to wolf predation are similar. In North America, where most observations have been made about wolf interactions with mountain sheep, the species involved are primarily Dall sheep, bighorn sheep, and Stone's sheep, a subspecies of Dall sheep.

Dall sheep and Stone's sheep possess thinner horns than do bighorns. Stone's sheep is similar to Dall sheep, but Dall sheep are white and live primarily in Alaska and

in the northern Yukon and Northwest Territories of Canada. Stone's sheep range farther south, in northern British Columbia and the southern Yukon Territory (Nichols 1978); they basically range from silver-gray through black with a conspicuous white rump, belly, and back of legs (Geist 1971). Bighorn sheep originally inhabited mountain ranges throughout western North America, including southern British Columbia and Alberta, but their current range is more restricted to wilder and/or protected areas (Wishart 1978). Bighorn sheep coat color varies from dark to light brown or grayish brown, with a lighter rump and backs of legs.

The weight of bighorn sheep also varies considerably by area, with adult rams weighing 58–143 kg and females 34–91 kg (Blood et al. 1970). Dall rams weigh 82–114 kg and Dall ewes, up to 64 kg, with Stone's sheep generally a bit larger (Nichols 1978). Newborn, mountain-sheep lambs weigh 3.4 kg (Wishart 1978; Nichols 1978). Bighorn, Dall, and Stone's sheep can live up to 16 yr or longer (Wishart 1978; Nichols 1978). Rams of all these three types of sheep possess rather massive horns that curl, with their sharp tips pointing outward as they mature to full-sized adults at about 3–6 yr (Blood et al. 1970). Ewes sport thin, only slightly curved but sharp, spikelike horns pointing backward that could penetrate an adversary approaching from the front if the ewe bent her head down.

Mountain sheep live in herds, with ewes, lambs, and yearlings forming one type of group and rams another. Living in groups no doubt confers on sheep all the antipredator advantages it does on other species (see chap. 1). Conceivably there might be occasions when group living is detrimental to sheep, however. This is because, when pursued by wolves, herds often seek promontories or steep precarious positions on mountainsides where wolves cannot reach them. Some such spots might be too small to fit the entire herd, leaving some members at the mercy of their attackers.

Mountain sheep possess keen senses. Under the right conditions, they can scent a human at 350 m and see one from 1 km (Geist 1971). Murie (1944) observed Dall sheep sensing a wolf some 800 m away (account 13). Regarding sheep hearing, Geist (1971, 13) indicated that "nothing is known definitively about their hearing," but Nichols (1978) thought their hearing was "very acute." As with all of the wolf's prey, sheep are highly vigilant, and those in smaller groups spend more time on alert (Horejsi 1976; Berger 1979). Sheep heart rates increase more when the animals are exposed to disturbances by wolflike stimuli than by other types of disturbance (MacArthur et al. 1979).

It is clear from the observations of wolves hunting sheep that the two species are well matched in speed, with prime, healthy sheep able to outrun or outlast wolves, especially when the sheep are in their element, running uphill or through rugged terrain (accounts 21, 24). Almost always, such fleeing sheep are striving to reach a ledge or pinnacle that wolves cannot. "The primary strategy of mountain sheep to escape predators is to place obstacles in their path. This demands that sheep stay close to appropriate obstacles and dash to these when predators approach. The most common obstacles chosen by sheep are steep, sharply dissected cliffs, where gravity and footing are in the sheep's favor and where they may dodge quickly out of the predators' sight. Ragged cliffs permit hiding, as does shrubbery" (Geist 1999, 201). Many of Murie's (1944) observations of Dall sheep in Denali Park (below) attest to the truth of this statement.

Even when caught short, away from escape terrain, however, mountain sheep still can defend themselves. For example, occasionally sheep leave one mountain and cross intervening flats to another. They usually do so in groups and tighten closely if approached by wolves. During an attack, sheep try to face wolves with their horns and butt them (account 29). Murie (1944, 102) cited a telling observation:

On one occasion Mr. Swisher said he let a sled dog chase some rams. They turned and faced the dog with lowered horns. After thus threatening the dog, the rams started up the slope, and when the dog followed, they again

turned on him. . . . A dog belonging to Joe Quigley, a miner, was said to have escaped one night from his camp in the sheep hills. The dog returned two or three days later, badly battered. Some time following this event, when the team was driven up to some sheep carcasses, this dog was not at all anxious to approach them. The inference was drawn that the bruises the dog had suffered during his absence had been administered by a sheep.

Sheep also can often outrun pursuing wolves even when not in escape terrain. An observation by Hoefs et al. (1986, 83) is telling in this respect: "The wolves had obviously pursued this band for more than 30 min in more or less 'rolling terrain' without distinct escape features. This observation revealed that sheep in good physical shape can outlast and outmaneuver two wolves for a considerable length of time" (account 24).

Once sheep reach a mountain slope, their speed, agility, sure-footedness, and endurance give them considerable advantage over wolves. Not only can sheep bound up steep, rocky slopes well ahead of wolves, but the nimble animals are truly at home in maneuvering in this steep terrain. For example, Geist (1971) once saw a captive bighorn ram jumping from a standing position through a hole in a wall some 2 m above. Such ability serves them well in reaching cliffs, pinnacles, promontories, precipices, and other areas where they can better ward off wolves (accounts 34–36; fig. 5.1).

The specialization of sheep to inhabit, and maneuver in, steep, extremely rugged, rocky terrain serves them well in the face of wolf predation. Thus, most strong, healthy sheep withstand such predation for most of their lives. According to Murie (1944, 110): "My observations indicate that weak animals are the ones most likely to be found in . . . vulnerable situations." Other support comes from Haber (1977), who watched wolves chase a total of 893 sheep and kill only 29 of them—a 3% success rate. Even when calculating success rate per sheep band chased, success was only 24% (Haber 1977). Haber's results are also in accord with earlier findings cited by Murie (1944, 103):

Charles Sheldon [1930, 315] tells of following the trails of two wolves in March and finding that on nine occasions they had chased sheep unsuccessfully. On eight of the chases they had descended on the sheep from above.

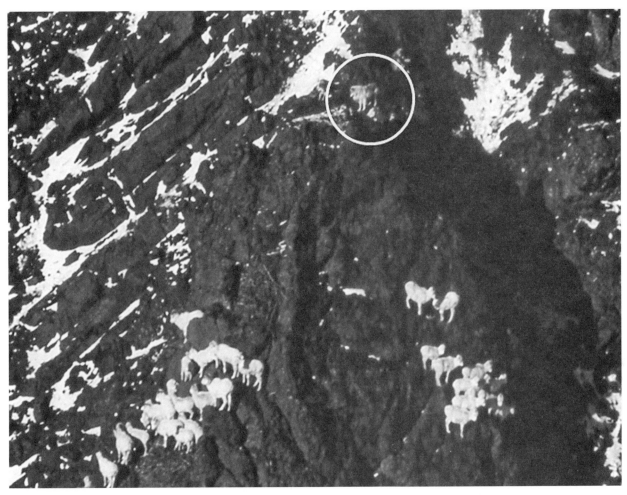

FIGURE 5.1. Dall sheep often escape wolves by taking refuge on pinnacles or precipices that wolves can't reach. (Photo by T. J. Meier [Mech et al. 1998].)

He said that the sheep in the region had become badly frightened, and that "most of them kept very high." The occurrence of so many unsuccessful hunts suggests that these two wolves were testing out each band, hoping eventually to find a weakened animal or to gain some advantage. It appears that wolves chase many sheep unsuccessfully and that their persistence weeds out the weaker ones.

Hunting Accounts

1. May 26, 1939—Denali Park, Alaska (Murie 1944)

Some observations made in the sheep hills bordering East Fork River show that, on a relatively steep smooth slope, sheep are easily able to avoid a single wolf. With a companion, I had climbed to the top of a ridge from which I had a view of some snow-free ridges on the other side of a small creek below me. I noticed a band of sheep resting on a smooth slope, slanting at about a 40° angle or less. While I watched, the sheep bunched up and ran off to one side about 30 m. Through the glasses I saw a gray wolf a short distance above. It loped toward them, and the band split in two, some going upward around the wolf, the others circling below it. When the wolf stopped, so did the sheep, only 30 or 40 m from it. The wolf galloped after the lower band, which ran downward and then circled, easily eluding the wolf. Compared to the sheep, the wolf appeared awkward. After a few more sallies the wolf lay down, with feet stretched out in front. One band lay down about 70 m above the wolf, the other about 50 m below. Only one sheep in the lower band faced the wolf; the others, as usual, faced in various directions. One sheep fed a little before lying down. The lower group consisted of five ewes, one yearling, and three rams. In the upper group were four ewes, four yearlings, one 2-yr-old, and two rams. All rested for 1 hr. Then the wolf again chased the lower band, which evaded it as before by running in a small circle around it. A flurry of snow then obscured my view. When it cleared a few minutes later, the wolf was disappearing in a draw, and the sheep were grouped on the ridge above it.

In a short time, the wolf reappeared and slowly worked its way down the ridge to the creek bottom. Nine ewes, each with a lamb, appeared on the ridge near the draw that the wolf had just left. The lambs were at the time only a week or so old, but still they apparently had been able to avoid the wolf. The utter lack of fear exhibited by these sheep is quite significant, indicating that a single wolf can easily be avoided on a slope.

2. June 19, 1939—Denali Park, Alaska (Murie 1944)

Twenty-two ewes and lambs were feeding among the cliffs a short distance above four resting wolves, one of which was lying only about 200 m away. The sheep had already become accustomed to the presence of the wolves, for they grazed unconcernedly. Their confidence was probably due to the proximity of exceptionally rugged cliffs to which they could quickly retreat should the wolves attack.

3. August 3, 1939—Denali Park, Alaska (Murie 1944)

A band of 20 sheep ran up the slope of a ridge bordering the East Fork River. A little later, a wolf climbed the slope, making slow progress. Twelve sheep watched from a point up the ridge, three from some rocks not far from where the wolf went over the ridgetop. Two eagles swooped at the wolf a number of times, continuing to do so after the wolf was out of my view so that its progress could be followed by watching the eagles. The sheep quickly resumed grazing. They had not moved far from the wolf but had watched to see what it was up to.

4. September 15, 1939—Denali Park, Alaska (Murie 1944)

Five wolves (the East Fork band) trotted along the road to Igloo Mountain, then climbed halfway up the slope, which was covered with several centimeters of snow, and followed a contour level; three were traveling loosely to-

gether, a little ahead of the other two. Sometimes they were strung out, 50 m or more apart. Generally, they trotted but occasionally broke into a spirited gallop as though overflowing with excess energy. Opposite me, they descended to a low point, and the two gray wolves that had brought up the rear dropped to the creek bottom. A black rounded a point and came on three rams, which it chased. The animals then disappeared, but in a few minutes the wolf returned, and the rams descended another ridge and crossed Igloo Creek. As they climbed a low ridge on Cathedral Mountain, they kept looking back. When crossing the creek bottom they had not been far from the two gray ones, which traveled 2 km up the creek and then returned to join the others.

After chasing the three rams, the black wolf joined another black and the dark gray female, and all moved up the slope. One of them stopped to howl, possibly a call to the two grays on the creek bottom. These two turned back about that time and later joined the others.

On a pinnacle of a high ridge stood a ewe peering down at the approaching wolves. She watched a long time but moved away while the hunters were still far. The wolves went out of view on the other side of this same ridge. Later, some ewes farther along the ridge looked steadfastly to the west toward the wolves, and beyond these ewes were three more gazing intently in the same direction. The sheep had fled to the highest points and were definitely cautious and concerned because of wolves, which seemed to be coursing over the hills hoping to surprise a sheep at a disadvantage. The day before, I had seen a lamb in this vicinity with a front leg injured so severely it was not used. Such a cripple would not last long if found by these wolves. The habit of cruising far in its hunting gives the wolf opportunity to find weak sheep over a large range and to come on undisturbed sheep, some of which it might find in a vulnerable location.

5. September 20, 1939—Denali Park, Alaska (Murie 1944)

A ewe, lamb, and yearling were killed on the road. The victims had been bedded down near a sharp corner. Four or five wolves had come around the corner, made a dash at the sheep and captured them before they had run more than a few meters. Several sheep killed by wolves were found on and beside the road at Polychrome Pass and at Igloo Creek.

6. September 28, 1939—Denali Park, Alaska (Murie 1944)

Mrs. John Howard of Lignite, Alaska, saw three sheep, at least one of which was a lamb, crossing a valley 2 km or more wide, between two mountain ranges. A little later she saw two of the sheep galloping up the long, gentle slope leading to the "outside" range. The lamb ran down the slope and disappeared in a draw. The two sheep continued to the ridge, but she did not see the lamb again. Later in the morning, she saw two black wolves near where the lamb had disappeared. One of them crossed the flat to the south and returned, followed by two pups. Mr. Howard found the lamb partly eaten and saw the wolves nearby. Sheep are especially vulnerable when crossing valleys, for wolves can outrun them on the flats.

7. October 7, 1939—Denali Park, Alaska (Murie 1944)

I saw the track of a single wolf that had crossed Igloo Creek and then had moved up the slope of Igloo Mountain. After following the creek about 2 km, I saw two ewes and two lambs on a spur of Igloo Mountain watching a black wolf curled up on a prominent knoll on the next ridge about 200 m to the west. The wolf's ears were turned toward me, so to not alarm him I walked until out of view as though I had not seen him. Then I doubled back close to the bank and ascended the draw toward him. But when I came near the knoll he was gone, and the sheep were lying down.

After backtracking the wolf, I deduced that its actions were as follows: after crossing a creek, it climbed partway up the mountain. It followed a trail along a contour at the edge of some spruces near where a wolf (perhaps the same one) had surprised a lamb among the spruces a few days before. The wolf crossed some draws and small spur ridges and arrived at a ridge on the other side of which the four sheep were feeding in a broad swale. The wolf climbed the ridge, out of sight of the sheep, for 50 m, so that it was slightly above them. It then advanced slowly until within 150 m of the two ewes and two lambs and then galloped downslope toward them. The latter had escaped to the next ridge from which they were watching the wolf when I first saw them. The wolf chased upslope after them a short distance and then continued west to the next ridge where it curled up in the snow. This time the sheep had the advantage and escaped. It is significant that they did

not run far beyond the wolf; apparently they were confident that they were safe since above them was much rugged country. The behavior of the sheep is definitely conditioned by the terrain they are in and their position in relation to the enemy. Approach a sheep from above, and it feels insecure and hurries away. A sheep on a flat is much more wary and timid than one in rugged country.

8. May 3, 1940—Denali Park, Alaska (Murie 1944)

I heard a wolf howling nearby and noticed three alert rams, each on a pinnacle, peering intently below them. They continued to watch for some time, evidently keeping an eye on the wolf. They apparently felt safe where they were even though the wolf was directly below them.

9. May 7, 1940—Denali Park, Alaska (Murie 1944)

A scattered band of sheep moved slowly from one ridge to an adjoining one. The movement was so definite and consistent that I suspected the sheep were moving away from danger. A little later I peered into the draw below the ridge first occupied by the sheep and saw a black wolf investigating some cleanly picked sheep bones. The sheep had simply preferred to have a ridge between them and the wolf.

10. August 3, 1940—Denali Park, Alaska (Murie 1944)

Dr. Ira N. Gabrielson and I watched a black wolf trot leisurely down a short draw on the ridge opposite us and descend the narrow stream bed bordered by the steep slopes. Two rams on the slope below watched the wolf, and when it trotted out of their sight they moved where they could again see it. Seven other rams grazing a short distance from the two paused but briefly to look. The wolf stopped a few times to look up at the rams, but continued on its way until finally hidden by a ridge. The rams and the wolf had shown definite interest in one another, but that was all. The wolf probably examines all sheep in the hope of discovering an opportunity for a successful hunt, and the sheep keep alert to the movements of the wolf so as not to be taken by surprise or at a disadvantage.

11. January 7, 1941—Denali Park, Alaska (Murie 1944)

The tracks of five or six wolves were seen along the road toward Savage Canyon. A short distance above the canyon, the wolves abruptly left the road and climbed a gentle slope to the top of an isolated, rocky promontory. They chased a lone ram down its steep side to the creek bottom where I found remnants of hide, as well as the stomach contents and the skull. Apparently the wolves had seen the ram on this isolated bluff and had turned aside to circle behind him and cut off his retreat to high ground. He was 12 yr old, past his prime, a weak animal. The method employed in capturing him—that of coming down from above and driving him down the slope—seems to be a typical hunting technique.

12. March 12, 1941—Denali Park, Alaska (Murie 1944)

Ranger Raymond McIntyre found the carcass of a recently killed 6-yr-old ewe at Toklat River. Tracks showed that two wolves had chased her down a steep slope and captured her at its base. The usual manner of hunting, coming down on the sheep from above, had been practiced. There was no indication of necrosis on the skull. Of course the sheep may have had some other weakness not apparent in the skull.

13. June 29, 1941—Denali Park, Alaska (Murie 1944)

About 60 ewes and lambs on the south side of a mountain all moved up the slope 100 m or more and stood with their attention centered on the terrain between us. Field glasses revealed a gray wolf loping west between me and the sheep, about 800 m from them. Some sheep soon began to feed; others watched until the wolf had passed. The day was dark and rainy so that the wolf, whose legs and lower sides had become wet from the brush, was unusually hard to see, yet the sheep had quickly discovered it.

Former Ranger Lee Swisher had seen six wolves suddenly come close to seven rams feeding out on some flats. The rams bunched up, and the wolves stopped 100 m away. They made no move toward the rams, which, still bunched, walked slowly and stiffly toward the cliffs. The slow gait was maintained until the cliffs were almost reached; then the rams broke into a gallop and quickly ascended the rocky slope. The incident seems to indicate that a wolf may to some extent recognize the ability of the rams to defend themselves. The wolves on this occasion may not have been hungry; possibly under other circumstances they at least would have made some attempt to single out one of the rams.

14. April 1970—Denali Park, Alaska (Haber 1977)

Five wolves detected 20–25 sheep (ewes, lambs, and one ram) from at least 2 km off and began a steady approach toward them along a series of open hills and steep ridges. There were five stops for intense socializing along the way, especially nose pushing, and after three of these one wolf (subordinate male twice, dominant male once) focused intently toward the sheep for 20–30 sec. The five wolves finally reached a point along the ridge just opposite the sheep; the sheep were about 45 m below the crest of the ridge on its steepest (south) side, grazing and loafing about, and the wolves were 7 m below the crest of the other (north) side.

The wolves then trotted up to the crest (heads just above it) and, for about 2 min, stood motionless, peering down intently on the unsuspecting sheep. At this point, the subordinate female backed slightly downslope and looped around toward an obvious escape route about 60 m west of the sheep. Just before the wolf reached this position, the others burst over the ridgetop and downslope at the sheep, fanning out and looping slightly toward their east flank, where there was another obvious escape route. The sheep were completely surprised, yet they reacted instantly and outdashed the wolves about 200 m to two secure, rocky areas, via the east-flank route. The wolves had apparently anticipated the two most likely escape routes and attempted to block them and trap the sheep between, but in this instance the sheep were too fast for them.

15. April 1971—Denali Park, Alaska (Haber 1977)

The 13 wolves encountered a 5-yr-old ewe in open, rolling terrain near the upper edge of a steep canyon. Based on tracks, it appeared that the wolves approached initially from the south and looped around to the ewe's east side, from where they fanned out and attacked. This forced the ewe to run to the canyon edge 100 m or so to her west, and she fell 6–8 m to overflow ice below. She was either killed outright in the fall or by the wolves a few minutes later after they descended via a slope 400–500 m to the north.

16. June 19, 1972—Alaska Mountain Range, Alaska (M. Groves in Heimer 1973)

At 2008 hr, a wolf came sprinting down a trail and into the mineral lick, scattering sheep in all directions, but mostly upward on five trails. I first caught sight of the wolf in a grassy area as it was sprinting. As the wolf "dove" into the lick, the sheep on the wolf's left ran out via two trails and between them. The sheep on the wolf's right started out a different trail. Some sheep ran horizontally in the lick and then turned up the lick at the foot of the bank and ran out along a trail. Some also ran around the side of the bank and out another trail. (All this happened when the wolf entered at 2008 hr.)

The wolf slowed a bit either to collect its footing or to decide which sheep to catch or both. It selected a lamb that was running around the top of a rock outcropping down along one side, then into rocks a short distance. The wolf was trailing approximately 25 m behind. At the moment the lamb ducked into the rocks, the wolf would not have been able to see it because the wolf was just reaching the top of the outcropping.

The wolf moved quickly down around one side of the outcropping and began hurriedly searching for the lamb below the outcropping, then rapidly extended its search through some little ravines of the creek bottom down to the edge of my view. On not immediately finding the lamb, the wolf began searching more slowly and thoroughly as it worked its way back up the creek bottom and along the side of the outcropping. It passed within 4 m of where the lamb was hiding in the rocks and continued searching up, down, and around, then up over the top of the outcropping. These proceedings covered about 14 min. I was watching the wolf most of the time, but whenever I looked, the lamb was holding a frozen "point" at the wolf.

The wolf continued away from the lamb. By this time, the wolf seemed to have given up seriously searching and was merely sniffing about as it moved on for about another 7 min. The wolf exited where it had entered, walked on out on a trail, and continued to a sidehill below the ridge until about 200 m up from the lick, then walked up over the ridge out of sight. (The wolf was seen from the valley bottom at this time heading south.) Sheep were visible in the area. At 2145 hr, 1 hr and 35 min later, the lamb was still in the rocks though the tension had eased. The lamb was lying down.

It is strange that I saw only six sheep in the lick when the wolf entered, because it looked like more than six when they scattered. Several sheep had just gone over the ridge from the lick before the wolf came, and as the wolf entered the grassy area, there seemed to be several sheep there also. I believe the wolf was chasing a band back down into the lick. There were two lambs present in this band, one of which was singled out as the intended victim.

At 2310 hr, the wolf showed up again in the creek bottom below the rocks where the lamb was still resting, sniffed, and looked around for 3 or 4 min, then departed. Both times, the wolf was there it looked toward the lamb several times without seeing it. At 0130 hr, the lamb was still in the rocks but was moving around on the outcropping quite a bit, and at 0253 hr, the lamb finally left the rocks.

17. July 20, 1977—Finlay Range of the Eastern Rocky Mountains, Central British Columbia, Canada (Child et al. 1978)

We observed a lone wolf interact with a band of Stone's sheep. The encounter was observed from approximately 800 m, using 8 × 30 binoculars and a 20 × 50 spotting scope. At 1743 hr, the wolf was above our campsite following a game trail along a glacial arête heading west. It stopped temporarily, surveyed our position and campsite for a few seconds, then continued to walk west following the height of land uphill toward an adjacent peak. At 1752 hr, the wolf was silhouetted against snow patches about 200 m from the summit but disappeared at the summit.

At 1805 hr, three ewes and three lambs suddenly appeared running directly downslope from the outcrops within which the wolf had disappeared. Shortly thereafter, the wolf was running down the same sharp inclines in hot pursuit of the sheep. It was 40 m above and behind the sheep and seemingly gaining on sheep below. Near the 1,370-m level, the sheep veered sharply and headed into the rocky ledges as if to seek an alternate route to escape. The wolf halted as abruptly, paused for a few seconds, surveyed the escape of the sheep below it, then reversed its direction and moved upslope as if to regain its advantage above the sheep. For 3 min, the wolf searched the area and occasionally looked downslope toward the sheep, perhaps attempting to detect the exact position of the sheep below. At 1820 hr, the wolf then moved downslope about 30 m and disappeared among the crags. Suddenly, it appeared chasing a single ewe and its lamb. At 1822 hr, the lamb tumbled 75 m from the rocky ledges onto the rock scree below. Within seconds, the ewe, in full gallop, fell as the lamb had over the same ledge. The wolf then approached the cliff, stopped momentarily, looked toward the sheep, then turned about face and moved upslope, apparently to seek easy access to the carcasses.

On our arrival we found not two, but three carcasses among the scree: the ewe, its lamb, and a wolf. From markings and general coloration, we concluded this to be the same wolf we had observed. We can only speculate that the wolf had also met its death by falling from the cliffs while attempting to gain access to the talus slopes and the sheep. At no time had we observed a second wolf. The next day we surveyed the area by helicopter, and since we saw no other wolf in the vicinity, and because the three carcasses were still in place, our conclusion as to the identity of this wolf seems substantiated. The wolf was a young female with no evidence of a recent litter.

18. January 5, 1982—Kluane National Park, Yukon Territory, Canada (Hoefs et al. 1986)

At 1140 hr, D. B. observed three wolves on the lower slopes of Sheep Mountain's west face. Temperature was -35° C and visibility poor because of drifting snow. The wolves came from the east, with one traveling higher upslope than the others. At 1155 hr a wolf noticed a sheep 100 m above. It gave chase, and the sheep immediately ran toward a rock outcrop 200 m uphill. After they reached it, the wolf gave up and joined the others.

19. May 1, 1982—Kluane National Park, Yukon Territory, Canada (Hoefs et al. 1986)

At 1000 hr, two wolves appeared from at an elevation of about 500 m above the valley floor. From 1000 hr to 1115 hr, they worked their way 2 km south and disappeared. At about 1330 hr, wolves were howling in the immediate vicinity. The closest band of sheep, consisting of two ewes, a lamb, and a young ram, were bedded nearby. The sheep suddenly jumped up and ran. Their behavior reflected surprise and panic, since they bumped into each other, and the lamb was knocked over. Two animals—a wolf and a sheep—were a few steps behind the others, and the last (a wolf) charged at the second-to-last and knocked it over. The fallen sheep regained its footing once, but the wolf was on it immediately and held it down with its front feet. The other sheep had reached

the safety of the canyon; the ewe caught was only 10 m from a drop-off.

The wolf bit the sheep repeatedly in the abdominal region while it struggled. Whenever the sheep lay quietly, the wolf looked at the observer—less than 70 m away in full view—and at the other sheep that had climbed out of the canyon and were watching only 50 m upslope. The young ram even bedded. The wolf would take a few steps toward the other sheep, and the downed ewe would kick again. The wolf would return to her and bite her in the back. It repeated this four times. The wolf finally bit the ewe in the lower back and shook her viciously, thereby pulling her about 5 m downhill toward the canyon. After that, she did not move again. The ewe was first downed at 1400 hr, and by 1420 hr, she appeared dead. The wolf stood over her for another 4 min, watching the observer and the other sheep but did feed. It then trotted uphill and disappeared.

The dead ewe was a 13-yr-old animal, marked during the 1970–71 winter. The tear into the abdomen was 10 cm long, but no organs had been torn out and there was very little bleeding. She was not pregnant, the viscera were intact, and there was no internal bleeding. The neck bite appeared to have been the fatal injury. Two fingers could be pushed into the holes created by the wolf's fangs and the splintered backbone was easy to notice. Analysis of femur bone marrow revealed that the ewe was in poor condition.

20. May 15, 1982—Kluane National Park, Yukon Territory, Canada (Hoefs et al. 1986)

At 1700 hr, M. Hoefs and H. Hoefs counted 12 ewes and yearlings and two young rams on an area 200 m wide and 100 m high at an elevation of 1,400 m. At 1703 hr, a wolf approached from the north along a sheep trail. The closest ewe was only 100 m away and grazing near that trail facing away from the wolf. The wolf approached at a trot, and some of the other band members began to run. The ewe did not notice until the wolf was about 20 m away. She then ran along the sheep trail, which followed the contour of the slope more or less horizontally, closely followed by the wolf for about 80 m. The wolf was gaining on her.

Instead of turning uphill like the other sheep, the ewe ran downhill 60 m into a depression covered by low bearberry shrubs at that elevation but willows and other taller shrubs farther down. The ewe was slowed by these shrubs, and the wolf caught up to her after about an 80 m downhill chase. We could not observe the struggling pair very well in the shrubs. While we approached the site, the wolf ran out once and made a short dash at two young rams that stood watching about 100 m uphill. However, the wolf gave up immediately and returned to the kill. It had not noticed our approach, but when we were about 50 m from the kill, the wolf ran uphill. The ewe was 13 yr old and her bone marrow was fatty. She was pregnant with a male lamb, almost ready for birth. The only injuries on the ewe were a bite in her lumbar region and a broken backbone.

21. May 17, 1982—Kluane National Park, Yukon Territory, Canada (Hoefs et al. 1986)

At 1240 hr, G. Calef, M. Broderick, M. Hoefs, and H. Hoefs were observing sheep run east and saw a wolf approaching from the west at an elevation of about 1,400 m. It was in no hurry and sat repeatedly. The wolf inspected a previous kill site and continued to hunt eastward. Only one ewe was left on this slope, grazing at a lower elevation about 30 m from a steep canyon. The wolf ran to within about 30 m of the ewe, stopped, and lay down, digging in and positioning its legs like a cat before the final jump. The ewe still had not noticed it. The wolf then stood and backed up, perhaps trying to approach the ewe from a higher elevation. The ewe noticed it, turned, and ran. The wolf pursued immediately but did not get close, losing much distance when crossing a steep canyon. For another 25 min, the wolf continued to hunt and got very close to a band of seven sheep. This band ran uphill first and then downhill, circling the wolf almost as if playing with it. The wolf repeatedly sat down and watched this band not more than 30 m away, and the sheep stood and watched it. The wolf finally left to the east about 1310 hr.

22. May 18, 1982—Kluane National Park, Yukon Territory, Canada (Hoefs et al. 1986)

At 1340 hr, M. Hoefs and H. Hoefs had located a ewe with a lamb in the cliffs above at an elevation of 1,900 m. She was standing on a sheep trail below steep cliffs and nuzzling a lamb, which was probably born the previous night. We had observed this pair through a spotting scope for about 5 min when a wolf suddenly appeared from above. It grabbed the lamb by a leg and dragged it about 10 m uphill. The ewe had hardly time to dash away a few meters. The ewe ran back and forth a few times and even toward the wolf once. At this time, the wolf rushed her but followed her only a few meters. It finally left with the lamb in its mouth.

23. July 3, 1982—Kluane National Park, Yukon Territory, Canada (Hoefs et al. 1986)

At 1340 hr, D. Burles observed a wolf traveling rapidly up a drainage gully that led onto a ridge where three rams were grazing. The wolf continued uphill a short distance and cut across the slope to another shallow drainage gully that led toward the top of the ridge and the sheep. When within 50 m and probably within view of the rams, it crouched and crept forward for 2–3 m and then dashed 30–40 m toward them. The wolf was among the sheep before they could react. It ran almost past one ram, then turned and jumped on his back. At that instant, the sheep turned and began running directly downslope. The wolf followed with a loose grip on the ram's back but lost control after 2–3 m and fell off, making a headlong somersault down the hill. The sheep continued to run downslope a short distance and cut across the slope toward a rock outcrop. The wolf lay motionless for 3–4 sec after its fall and followed slowly after the sheep. The other two sheep in the group began to run upslope when they became aware of the wolf in their midst but stopped after a short distance to watch the chase below them. They, too, headed toward the rock outcrop after the initial chase.

24. August 3, 1982—Kluane National Park, Yukon Territory, Canada (Hoefs et al. 1986)

At 1905 hr, we observed a band of seven ewes and one lamb about 1 km away on another ridge east of us. We had observed this band for about 10 min when two wolves appeared from above to within 30 m of the sheep before being noticed by the sheep. The sheep then ran downhill, closely followed by the wolves, which fell behind immediately. The only ewe that had a lamb climbed into a small rock outcrop, and the wolves ran by her not more than 5 m away following the other sheep. After a 200 m downhill pursuit, the sheep and wolves disappeared. At about 1935 hr, we saw six adult sheep running up the mountain west of us. Shortly after, a wolf followed them, more than 200 m behind. The sheep ran up the mountain with apparent ease; the wolf was obviously tired and gave up after a few minutes. The wolf then returned to our ridge and joined the other wolf, which had bedded without being noticed by us. The wolves had obviously pursued this band for more than 30 min in more or less "rolling terrain" without distinct escape features. This observation revealed that sheep in good physical shape can outlast and outmaneuver two wolves for a considerable period.

25. February 21, 1985—Kluane Region, Yukon Territory, Canada (Sumanik 1987)

Four members of the Tepee Lake Pack were traversing the upper elevations of a mountain. All four were separate. Tracks on the south slope indicated there had been much wolf activity. One wolf suddenly left the ridge and bounded down the south aspect, generating a small avalanche of snow just ahead, similar to a boat's bow wave. Any sheep in the avalanche path would have been knocked down. Whether this wolf was taking advantage of the snow conditions to improve hunting success or whether sliding snow occurred incidentally to this animal's downhill charge is unknown. The wolf ran almost to the mountain base, passing many small, rock outcrops, clearly hoping to surprise and overtake a sheep among the outcrops. Some 5-min later, another pack member ran down the north aspect similarly. However, at some points a very hard crust caused the wolf to lose balance and traction.

The other two wolves, although separate from each other, were pursuing a band of 10 nursery sheep on the ridge. The tightly bunched herd stopped occasionally to observe the trotting wolves 200 m behind. The sheep left the main ridge and ran to the end of a perpendicular ridge that ended with a sheer cliff at least 200 m high. The sheep remained tightly packed at the cliff's edge. Both wolves approached to within 100 m, but one soon descended the mountain. The other stood assessing the situation but also descended soon after. Although any attack on the sheep in their position would have been extremely dangerous, an approach to within 5 m was very feasible, and I was surprised that the wolves did not approach in hopes that a sheep would panic and flee the cliff edge. Shortly after the wolves left, the sheep returned to the main ridge.

26. March 14, 1985—Kluane Region, Yukon Territory, Canada (Sumanik 1987)

While I located the Cement Creek pair of wolves, three mature rams were traversing a 30 m high cliff above a tributary of Cement Creek. Female Wolf 1035 was paralleling the sheep's course from the cliff top. The two leading sheep were slightly ahead, and they soon outdistanced the wolf. However, the third sheep stopped abruptly and faced uphill toward the wolf. As it did so, snow and rock fell away from beneath its hind feet, and for a brief moment it appeared that the sheep would fall to the creek ice below. The ram recovered, adjusted his footing, and watched Wolf 1035, while she looked down from above. The female wolf started down the cliff toward the sheep but retreated to the cliff top when she could not reach the ram and footing became unsafe.

We left the scene once this standoff had continued for several minutes. During the encounter, the male wolf (later known as 1065) was 20 m upstream, also at the cliff edge, lying among a willow bush. During the entire hunt he never moved from the willows, and it was difficult to say whether he was sleeping or hiding.

27. March 19 1985—Kluane Region, Yukon Territory, Canada (Sumanik 1987)

The Cement Creek wolf pair were resting on a mountainside 500 m above the valley floor. As we approached, female Wolf 1035 arose and walked west on the same contour. Four hundred meters above was a lone, half-curl ram. The ram was looking at the wolf, and the wolf would occasionally look up at the sheep. When 1035 reached the ridge that the sheep was on, the wolf started uphill toward the ram. As 1035 approached, the sheep began to walk uphill, directly away from the wolf. When the ram reached the mountaintop he walked west. Once 1035 reached the mountaintop, she trotted after the ram and only when the ram traveled out of sight, behind a rolling knoll and below 1035, did 1035 pursue aggressively. The wolf approached the ram at high speed, but when she was 10 m away, the running ram gained ground, darted down the mountain, and then contoured east just below the crest of the mountain. The wolf, easily outdistanced, gave up.

Meanwhile, the male wolf, oblivious to the activity above, stood up and walked directly uphill on a course intersecting that of the fleeing ram. The ram passed just

10 m above the wolf, and the seemingly disinterested wolf did not pursue the sheep. I believe the ram and the male wolf knew nothing of each other's proximity until the sheep passed by. The wolves reunited and gave up the hunt.

28. March 19, 1985—Kluane Region, Yukon Territory, Canada (Sumanik 1987)

Three members of the Generc River Pack were harassing a band of 20 nursery sheep on a rolling, talus ridge. The wolves had surrounded the sheep, and rather than gradually tighten the circle, the wolves pursued the sheep individually. There was no escape terrain nearby, and a kill appeared imminent. One wolf was above the herd, and two below. Each wolf took turns pursuing the herd. Once one of the lower wolves managed to split the herd, but both sheep herds remained tightly packed. One herd climbed a rock outcrop while the other stood on the open talus waiting for the wolves to actively pursue again. The sheep on the outcrop seemed especially vulnerable, and I expected the wolves to surround the outcrop and attack. Instead, they discontinued the hunt and walked away. I felt as though the wolves could easily have made a kill, especially if all eight members had been present. The hunt lasted less than 5 min.

29. March 25, 1985—Kluane Region, Yukon Territory, Canada (Sumanik 1987)

The Cement Creek pair were hunting again, similarly to the March 14 hunt in Cement Creek. Female Wolf 1035 was at the top of a 300 m tall, clay cliff overlooking the Klutlan Glacier. Again she was paralleling a three-quarters-curl ram, which was traversing the cliff 20 m below. The ram stopped and turned uphill, almost losing balance. For a few moments they stared at one another, and then 1035 walked down to within a half meter of the ram. The ram tried to butt the wolf, and contact was very close. Wolf 1035 turned and returned to the cliff top. We left after what appeared to be a standoff. During the entire period, the male wolf (1065) lay at the cliff edge. It was difficult to tell whether he was resting or lying in wait.

When we returned the following morning, the ram was standing in the same location. The wolves were sleeping 400 m from the cliff edge. The ram walked up to the cliff top, peered over briefly and dashed to the closest mountainside. The wolves continued sleeping.

The almost exact similarity between the first and third hunts by the Cement Creek pair is likely not a coincidence. The inactivity of the male in all three hunts could reflect a differing tactical hunting style or role. The differing behavior by each wolf may have reflected their differing nutritional status. Female 1035 was in very poor condition, while 1065 was in excellent physical condition.

It appeared on all occasions that a unified approach, with a more active, cooperative role by the male, would have enhanced hunting success. However, I may be overlooking a very logical reason for their strategy.

30. March 23, 1986—Kluane Region, Yukon Territory, Canada (Sumanik 1987)

The Donjek Pack was descending a long, rolling ridge fanned out across the ridge when suddenly they all sprinted downhill for 1 km. Although many sheep were on adjacent ridges, none was on this ridge. The strategy appeared similar to the Tepee Lake Pack, where they hoped to surprise and overtake a sheep among the outcrops and undulating terrain. One could argue that the wolves were merely running, but their behavior before the run indicated they were searching for prey.

31. April 10, 1986—Denali Park, Alaska (J. Burch)

At 1441 hr, three wolves, including 215 and 221, were chasing two sheep. There were four other wolves in the area that did not participate. One sheep was a full-curl ram, the other unknown. The larger ram turned and went up a steep slope and was not pursued. The other continued down a side slope then up the other side; when the three wolves got to the bottom of the side slope, two abandoned the chase and one continued. The sheep and wolf climbed the slope and into dark shadows—we lost sight of them. The wolves never came closer than 200–300 m. The other wolves were just milling about and traveling slowly.

32. January 26, 1987—Denali Park, Alaska (J. Burch)

Six wolves, including 217 and 221, all spread out slowly and approached three sheep; the sheep appeared to have detected the wolves already. The wolves loped up the ridge halfheartedly, and the three sheep fled. The wolves stopped running and milled about. The wolves did not appear very interested in the sheep. The wolves were just leaving a moose kill.

33. March 1, 1987—Denali Park, Alaska (J. Burch)

At 1405 hr, two wolves, including 231, made a halfhearted chase of one sheep on a steep, gravel slope for about 30 sec, but only got to within 200 m. The wolves had already killed one sheep and consumed about half of it. This carcass was downhill south of the wolves and the chase by about 300–400 m.

34. March 6, 1987—Denali Park, Alaska (T. Meier)

At 1408 hr, six wolves, including 217 and 221, had a band of 40–50 sheep cornered on a rock pinnacle. Two wolves were around the bottom, and one on top, sometimes within 30 cm of a sheep. Periodically, one or several sheep would make a break for another pinnacle nearby or for the mountain slope. The wolves would pursue but caught none. At 1841 hr, the wolves were 2.5 km away, traveling away from the site, and no sheep was at the site.

35. November 22, 1987—Denali Park, Alaska (M. Britten)

At 1230 hr, seven wolves were attacking a band of about 25 sheep on an alluvial fan. About three wolves, including 217, actually were in with the sheep, and I thought that the wolves would easily catch one. The sheep made it to an outcrop, where the wolves quit the chase but sat for a few minutes then left.

36. February 27, 1988—Denali Park, Alaska (T. Meier et al.)

At 0902 hr, at least four wolves including 980 were after a couple of sheep and had one young ram cornered on an isolated outcrop high above Moody Creek—wolves able to move above, below, and beside a safe area. The ram several times moved to another spot, and wolves came within 1 m and would nearly catch it. Situation remained the same through three visits with airplane.

37. March 4, 1988—Denali Park, Alaska (T. Meier)

At 0940 hr, six wolves, including 980, were sleeping near a rock outcrop where several ewes and lambs were at bay. The wolves awoke and approached the sheep; one lamb backed off and fled across a snow slope and down a gully. A wolf pursued and caught it, then descended 60 m before stopping. Three other wolves arrived and began feeding. Wolf 980 lagged behind. The carcass slid another 30 m before stopping again. All five males were feeding on separate parts of the carcass within a few minutes.

38. June 28, 1988—Denali Park, Alaska (T. Dawson)

At 1030 hr, we observed from the ground a lone wolf stalking, chasing, and killing a spring Dall sheep lamb on the steep, rocky hillside on the southeast side of a creek. We had followed the wolf as it hunted arctic ground squirrels from the top of a pass along a road. We had seen sheep grazing very low as we passed the area earlier. The sheep were grazing 100 m above the creek in an area of small rock outcrops that made a diagonal line across the hillside—high end to the left, low end to the right near the creek. Near the upper end of this series of outcrops were clusters of more serious escape habitat within 30 m of the grazing sheep. The wolf scented the sheep as soon as it crossed to the southeast side of creek. It moved more stealthily through the streamside willows until it saw a few of the ewes and lambs grazing above.

A few of the nearest sheep had seen something suspicious in the willows below and began moving into a hillside depression near the top of the diagonal line of outcrops. The wolf froze until the sheep disappeared, then sprinted directly to where they were last visible.

Streaking over the lip of the depression, its momentum carried it nearly into the band of sheep, scattering them. Ewes and lambs frantically scrambled to escape into rougher terrain or straight uphill over the ridgetop. The wolf pursued two ewe-lamb pairs that contoured to the left (northeast) around the hill into jagged rock. Perched in the center of this very small island of escape rock, the sheep fidgeted nervously. The wolf circled the rock. First attempts to reach the sheep from below were fruitless, so it circled above. There it found precarious footing along a knifelike projection of rock, eased out to a point about 1.5 m above the sheep and peered over.

At the sudden appearance of the wolf's head, the sheep panicked, leaping downhill to the grassy slope below the rocks. Leaping easily 4–5 m vertically downward, the wolf sailed off in pursuit. Landing just behind one ewe-lamb pair, his close presence frightened the lamb into choosing an escape course that separated it from the ewe. It tried to evade the wolf by running randomly back and forth across the least severe of the escape rocks instead of following its mother uphill.

After two or three zigzag passes across the face of the

hill, the wolf had closed to within a meter of the running lamb. The wolf managed to gain ground as the pair raced across some very rough footing. As the wolf closed, it appeared that the lamb was tiring, and soon it turned diagonally downhill. The wolf seemed to throw itself into the last bounds of the chase and quickly overhauled the exhausted lamb.

Surging ahead, the wolf grabbed the lamb by the right stifle (i.e., knee), and the two tumbled down a short scree slope. In the brief struggle, the wolf changed grips from the lamb's rear leg to its throat and held on until the lamb lay motionless. Panting, the wolf rested next to the lamb for 2–3 min. When the lamb began waving a foreleg weakly, the wolf returned, biting it quickly several times in the anterio-ventral thorax. It then returned to its resting spot. The lamb convulsed once or twice and then was still. After resting 20 min, the wolf began feeding. It quickly (20–30 min) devoured most of the hindquarters and one front quarter and cached the remains.

39. September 27, 1988—Denali Park, Alaska (T. Meier)

At 1030 hr, two wolves were seen, including 217. One was chasing two sheep; the wolf was about 200 m behind at first but caught up with the rearmost sheep and cornered it on a small knoll with no escape terrain. But the wolf immediately broke off and continued on the trail of the front sheep. A second wolf pursued a sheep down a brushy slope. The two wolves caught the sheep on the creek bed and killed it there. Both sheep were chased across a slope and then downhill—some very limited escape terrain was available above the kill site.

40. November 10, 1988—Denali Park, Alaska (T. Meier)

At 1535 hr, we saw two wolves (probably 251 and 307) chasing two ewes and two lambs up over 5,020 peak. They went southwest, then northwest along ridgetops, mostly uphill. The darker wolf persisted for around 1.6 km but never came closer than 100 m. The sheep had clear running on ridgetops, but the wolf continued. The lighter wolf had given up sooner. Five pups were 1.6 km and 300–450 m above and behind. The pursuit started when we saw it—because we saw the sheep standing still on 5,020 peak, and we had been in the area for 10 min looking at the five pups.

41. February 15, 1989—Denali Park, Alaska (T. Meier)

At 1200 hr, seven wolves including 5051 and 217 had cornered seven sheep on some rocks down near the bed of a river. There were five wolves in one spot and two alone on ledges. All were concentrating on the most vulnerable sheep and had it surrounded but could not get to it. Most of the sheep were large rams; the one the wolves were after was young or a ewe.

42. February 23, 1989—Denali Park, Alaska (T. Meier)

At 0945 hr, four wolves including 341 and 343 had cornered around 40 sheep on a small rocky knoll near a creek. At 1623 hr, the wolves were no longer there but had cornered nine rams on another creek.

43. April 14, 1989—Denali Park, Alaska (J. Burch)

At 1200 hr, we located at least eight wolves including 5051 and 217 feeding on one sheep; five to six wolves had a lamb cornered on a rock pinnacle. A white wolf initially below the lamb climbed around to the top of the pinnacle while a gray shimmied out to within 30 cm of a lamb on sloping rock, but the wolf could not reach the lamb without falling. The white wolf (breeding female?) looked over, shimmied out on the rock, made contact with lamb but free-fell about 15 m (measured later) and bounced twice on rocks below and slowly walked away with head down. In the meantime the rest of the wolves (five) grabbed the lamb and fed. In 3 min, the lamb was 65% eaten with all the wolves off chewing on their own pieces.

44. June 29, 1989—Denali Park, Alaska (P. Del Vecchio)

At 1900 hr, about 10 scattered sheep were grazing on a bench near a glacier. Suddenly, a wolf (evenly gray) came running among them. It seemed to overshoot. The sheep moved to the edge of a cliff (which they were above). Other sheep on the bench were unaware of the wolf—they continued grazing. The wolf circled, doubled back, and weaved around behind hills to get closer to the sheep. All the sheep began walking to the cliff edges and followed the cliff edges up a drainage. The sheep crossed the drainage, and the wolf followed, but we left.

45. June 26, 1990—Denali Park, Alaska (E. Smith)
At 1835 hr, we watched from the ground a gray wolf chasing a sheep up a cliff. The wolf climbed after the sheep for awhile then slipped and seemed to give up.

46. October 8, 1991—Denali Park, Alaska (J. Burch)
At 1320 hr, we located 10 wolves, including 252 and 307, and all but two were sleeping or resting; two were just below them on a cliff face trying to approach a ewe and got to within 3 m and gave up (cliff too steep). The ewe remained motionless on a small ledge the whole time. 4 hr later, nothing had changed; the wolves were all asleep and the ewe was in the same place.

47. November 18, 1991—Denali Park, Alaska (J. Burch)
At 0929 hr, eight wolves, including 251 and 307, had 13 rams, ewes, and lambs cornered on an isolated rock outcrop. The airplane scared all the sheep off the rocks; the wolves arose and ran toward them splitting the group into eight and five. All the wolves pursued the eight back to the rock, coming to within 3 m of them and seemingly ignoring the five running away down the ridge without any escape terrain.

48. March 30, 2007—Jasper Park, Alberta, Canada (D. Dekker)
On the open hillside, an alert band of 26 ewes and their progeny were watching a black wolf walking along the rim of the escarpment, some 80 m below the sheep. Suddenly, the wolf sprinted up the steep incline. The group split into two, and part of one group veered downhill, bypassing the charging wolf, which turned and sprang after them in vain. After three similar attack sequences, the entire band had managed to get by the wolf and reach the safety of the canyon cliffs just in time.

49. March 29, 2012—Yellowstone Park, Wyoming (J. Tasch, C. Anton, H. Martin, M. Metz, and D. Smith)
A group of 15 bighorn sheep on a ridge were loosely together as seven Lamar Canyon wolves moved uphill, still 400 m away. When the wolves reached the top, they traveled to where the sheep were then bunched tightly. The wolves stopped on a cliff above the sheep and watched. As the wolves approached the sheep, the sheep fled downslope. One 11-mo-old pup pursued the sheep, while the rest of the pack remained spread out above. The sheep split into two distinct groups, which ran to adjacent cliffs and bunched together facing the wolf. The wolf stood between the two sheep groups while the pack still watched from above. After a minute, the wolf began to retreat but continued to stop and look back at the sheep. A sheep left the one group and ran to the other. The wolf saw this and immediately moved back toward the sheep. The wolf sat between the groups again as a second sheep ran from one group to the other. The wolf ran forward. Two sheep approached the wolf, and it stepped back. After another minute, the wolf retreated, and the rest of the pack left as well.

Throughout the entire 5-min encounter it was clear that the sheep had situated themselves in steep terrain where they were comfortable and the wolf pup was not. From our vantage point (valley floor), it appeared that the wolf could have advanced farther onto the cliff where the sheep stood, but the wolf made no such attempt. Wolf and sheep watched each other while fairly close, but the wolf never attempted to move closer. It was also interesting that no adult attempted to approach the sheep.

Mountain goats

Similar to mountain sheep, mountain goats inhabit steep rugged terrain, but goats are less common and occur more in the steepest mountains, primarily in the western United States and Canada as well as in south and southeastern Alaska. In Eurasia, several counterpart species, such as chamois, takin, tahr, and ibex, inhabit similar habitat in wolf range. However, the only descriptions available of wolves hunting any of these species are of mountain goats. Like sheep, mountain goats use the steep terrain that is their natural habitat to escape wolf attacks, especially sheer promontories, which they can reach much more easily than wolves. When caught short, goats can flee more effectively than wolves through their rugged terrain and, if necessary, at least some will try to run off a wolf (account 5).

Hunting Accounts
1. June 17, 1977—Herbert Glacier, Alaska (Fox and Streveler 1986)
At 1030 hr, an adult female wolf was observed at 650 m elevation climbing a 40° subalpine slope toward several goats.

The wolf came within 30 m of a nanny and kid watching its approach. The goats were only 10 m from very steep and broken terrain; they did not move but watched attentively as the wolf continued past them. A nanny and kid on relatively gentle terrain 40 m higher broke into a fast run for 30 m when the wolf suddenly appeared 50 m away. At the edge of steep and broken terrain, they stopped and watched the wolf continue upslope. The wolf then passed within 75 m of a nanny and one kid in steep, rocky terrain; they watched the wolf but did not move. It made no move toward any goats and continued upslope and disappeared over the alpine ridgetop at 1,100 m.

2. January 22, 1979—Herbert Glacier, Alaska (Fox and Streveler 1986)

At 1630 hr, four adult wolves were traversing single file across a predominantly snow-covered, 35° alpine slope at 800 m. Six adult goats and one yearling were feeding on the slope around a promontory such that neither group could see the other until 30 m apart. The wolves' line of travel took them directly toward the goats. When they saw the wolves, all the goats ran quickly to a steep rock outcrop 35 m in the opposite direction from the wolves. The goats split into two groups, one of three and one of four, with each stopping within the rock outcrop, bunching up and backing against vertical rock faces that prevented approach from the rear. They stood and watched as the wolves, already at the outcrop, split up, with two circling around above it and two waiting below. The wolves watched the goats for 1.5 hr until dusk but never confronted them. The next morning, the four goats had traveled 800 m to another rock outcropping while the three had moved over 2 km to an extensive area of steep rock outcroppings.

3. June 26, 1977—Dixon Harbor, Alaska (Fox and Streveler 1986)

At 1300 hr, an adult wolf was traveling along an alpine ridgetop and into broken rock outcroppings near the ridgetop at 700 m. An hour later, 18 goats came up over a smooth portion of the ridgetop 40 m from the wolf in the rocks. The nearest steep, rocky terrain, other than that around the wolf, was 350 m away. The wolf darted out of the rocks and into the group of goats, seized a kid by its upper rear flanks, shook it, quickly switched its grip to a point above the goat's shoulders, and trotted

back toward the rocks before the goats appeared to realize it. At that point, the 17 remaining goats ran full speed away from the wolf for 80 m, whereupon they stopped and watched the wolf disappear. The wolf carried the kid 200 m to a snowbank near the ridgetop. The goats stopped and began to feed again within 15 min.

4. September 24, 1981—Boca de Quadra, Alaska (Smith 1983)

A group of two adult females, two kids, and one subadult goats (sex unknown) were bedded in an area of broken rock interspersed with alpine vegetation on a 15°–20° slope at 1,100-m elevation 20 m below the ridgetop. A second group, two adult females and two kids, were bedded on a smooth, 30°–35° slope of alpine vegetation 100 m below and 100–150 m south of the other group. About 50 m farther south was a steep rock outcrop slope 45°–50° approximately 30 m wide and 10–20 m high.

At 0942 hr, a single adult wolf trotted over the crest of the ridge downwind and 50 m south of the first group. The instant the wolf appeared, the upper group bolted north along the ridge, remaining below the ridge crest in rocky terrain. The wolf chased the goats, but after 100–150 m it stopped. All five goats continued to flee at a panicked gallop 800–900 m upslope before topping the ridge out of sight onto an extremely sheer rock face.

In contrast to the first goats, the lower group rose quickly and walked at a deliberate but unhurried pace to the nearby outcrop. One pair climbed onto a narrow ledge, 5 m above the base and near the middle of the outcrop, making an approach by the wolf difficult or impossible. The other pair moved to the base of the outcrop and backed against a vertical rock face, thus preventing any flanking approach.

The different reactions of these two goat groups may have been due to their relative positions when the wolf appeared. The upper goats could have been intercepted by the wolf if they ran toward the rock outcrop. Their response was to flee to the nearest steep terrain almost 900 m away. The lower group was able to reach escape terrain before the wolf could attack.

5. August 30, 1995—west-central Alberta, Canada (Cote et al. 1997)

A herd of 40 goats (38 were marked) and 12 kids were foraging in an open slope at 2,010 m, 100 m from timberline when at 1255 hr two adult wolves (one gray and one black) ran from the forest and chased the goats uphill for 300 m to the closest rocky cliff. The wolves did not get closer than 40–50 m before the goats reached the cliff. At 1302 hr, the gray wolf approached the goats at the bottom of the cliff and, after a few attempts, grabbed a 3-mo-old, 23-kg, male kid. As soon as the wolf pulled the kid down the rocky ledge, his mother, a 7-yr-old, jumped down and charged the wolf. She hit it twice on the rump and missed it once. The wolf released the kid, and both mother and kid fled to the other goats on the cliff. Three other adult goats then charged the wolf and forced it to retreat. The wolf, apparently uninjured, returned to the other wolf 150 m away. The goats then disappeared to the other side of the escape terrain followed at about 200 m by the two wolves that skirted around the cliff.

At 1615 hr, the group of goats returned to feed on the same slope they had used in the early afternoon. At 1716 hr, the same gray wolf appeared alone at the ridgetop and started pursuing the goats that ran toward a rocky cliff. As the wolf approached the base, the last three goats changed direction and started to run toward the forest. The wolf caught up to the goats and grabbed the smallest one (a marked yearling female) by a hind leg, but the goat escaped and kept running toward the forest. The wolf recaptured the goat by the same hind leg while running downhill, and they rolled together 15 m downslope. The goat arose again but was quickly caught at the throat and knocked down by the wolf. The goat managed to stand and escape once again but was again recaptured, bitten at the throat, and died in less than 3 min. The wolf then disappeared in the forest (less than 20 m away) for 5 min. It came back to the carcass at 1736 hr and dragged it into the forest out of sight. At 1744 hr, the other goats started to bed in the cliff. Goat 75 (the mother of the dead yearling) looked for several minutes at where the wolf had disappeared and was the last goat to bed. She had not attempted to defend the yearling.

6. July 11, 1995—west central Alberta, Canada (Cote et al. 1997)

An adult wolf attacked a group of 63 goats, including 16 kids, feeding in an open forest at 1920 m but was unsuccessful.

7. August 20, 1995—West Central Alberta, Canada (Cote et al. 1997)

A juvenile wolf chased a group of 84 goats, including 20 kids, foraging at 400 m from a steep, rock face, but the goats ran to the cliff and the wolf never got closer than 30 m.

Conclusion

Murie's (1944, 109–10) generalizations of many years ago cannot be improved on in summarizing how wolves succeed in overcoming mountain sheep specialization and defenses:

> It is my impression that the wolves course over the hills in search of vulnerable animals. Many bands seem to be chased, given a trial, and if no advantage is gained or weak animals discovered, the wolves travel on to chase other bands until an advantage can be seized. The sheep may be vulnerable because of their poor physical condition, due to old age, disease, or winter hardships. Sheep in their first year also seem to be especially susceptible to the rigors of winter. The animals may be vulnerable because of the situation in which they are surprised. If discovered out on the flats the sheep may be overtaken before gaining safety in the cliffs. If weak animals were in the band, their speed and endurance would be less than that of the strong and they would naturally be the first victims.

The above hunting accounts of wolves hunting both sheep and goats conform to these generalizations but provide additional details.

Murie (1944) supported his conclusions about the wolves ending up catching more vulnerable individuals by examining the age, sex, and condition of Dall sheep in Denali Park based on examining the skulls of 829 sheep. His classic conclusion that only 5%–12% of the sheep remains represented "sheep in their prime which were healthy so far as known" (Murie 1944, 124) has been well supported to one extent or the other with wolves preying on all other major, wolf prey species (summarized by Mech and Peterson 2003).

Although the wolf's physical abilities cannot come close to matching the sheep's or goat's prowess in negotiating steep, rugged terrain, wolves can exploit weaknesses in sheep behavior, condition, or circumstances. Furthermore, there is every reason to believe that wolves can learn specialized methods to exploit those weaknesses. One such method is to try to get above their prey and charge downhill toward them, surprising them before they can reach escape terrain (accounts 14, 17, 24–26).

With quarry frequenting such rugged and varied terrain as mountainsides, there is much opportunity for chance to play a greater role in wolf hunting behavior than in wolf interactions with most other prey. Even the unevenness of some high-arctic tundra can play such a role, although a lesser one, with wolves preying on musk oxen (chap. 7), and the basic concept of habitat, terrain, weather, and so on randomly influencing wolf hunting behavior probably applies to wolves hunting every species. With mountain sheep and goats, the influence of such factors is just more obvious to humans. Thus, wolves coursing along a mountainside and suddenly encountering sheep around a sharp corner can surprise the sheep, and the wolves can possibly pick off a lamb or an older ram whose senses might not be so keen or which might just be caught too much by surprise.

Murie (1944, 108–9) recounted how a road cutting along a mountainside can artificially skew chance in the wolf's favor:

The highway favors the wolves in three ways. First, it gives them an easy trail along the entire winter range so that they can move more readily from one part of it to another. Second, it gives the wolves easy access into the cliffs themselves; they need not make a laborious climb to get among the sheep but can follow a smooth easy grade. Finally, each blind corner in the road—and there are many of them—is a hazard, for the sudden appearance of a wolf may give the sheep no time to escape. This is especially true when the sheep are bedded down on the road.

The road, being an extreme, artificial intrusion through sheep habitat, dramatically illustrates the role that varying topographic differences can make to wolf-sheep interactions, but one can also easily envision similar, natural topographic variations that can similarly affect wolf predation on sheep.

When all the above factors—the wolf's basic physical traits, its ability to learn, the varied terrain sheep and goats occupy, and sheer chance—combine with the natural variation in prey age, health, and condition, the net result is similar to that of wolf interactions with other prey: to succeed, wolves must make many attempts to capture their quarry. On average, they catch only a single sheep for each 33 that they chase (Haber 1977), and generally they end up with those not quite so quick and able as healthy, prime individuals.

6

Bison

WOLVES AND BISON share a special similarity. They are the largest living representatives of the suite of pursuit predators and hoofed grazers that once inhabited the vast ice age grasslands of North America more than 10,000 yr ago (Lott 2002). Wolves were the smallest in a diverse cast of large carnivores that included American lions, saber-toothed cats, short-faced bears, and dire wolves, a relative of the gray wolf and the largest canine before or since the Pleistocene (Wang and Tedford 2008). Bison shared their range with mammoths, mastodons, horses, and camels. The interaction between wolves and bison (fig. 6.1) was therefore only one, and arguably the least spectacular, of many charismatic predator-prey relationships that typified the North American plains during the Pleistocene. Nevertheless, it was the only one to survive the mass extinction at the end of the epoch, an event that was apparently triggered by large-scale environmental changes (Shapiro et al. 2004; Guthrie 2006).

Aside from being the largest predator-prey pair to survive the Pleistocene extinction in North America, wolves and bison also coexisted over the largest area. Their range extended from subarctic Canada to northern Mexico and across the United States, from the Great Basin to the Atlantic Coast. But by the beginning of the 20th century, this geographically widespread predator-prey system had largely vanished from the continent. European settlers had slaughtered the bison in the late 1800s, mainly for hides in high demand for industrial machine belts because of their strength and wear resistance. Wolves were trapped, shot, and poisoned to protect the livestock that replaced the bison.

Fortunately, two populations of wild bison survived, and each later became the focus of modern research on wolf-bison interactions. These bison persisted because they inhabited the most remote corners of their range.

The largest population inhabited the area that is now Wood Buffalo National Park (WBNP), Canada, and it reached a low of around 250 individuals by 1900 (Soper 1941). The smallest population survived in what is now Yellowstone National Park (YNP), and it declined to an estimated 25 individuals in 1902 (Meagher 1973). Whereas wolves persisted in northern Canada, including WBNP, throughout the 20th century they were extirpated from YNP by 1930 (Schullery 2004) but reintroduced 65 yr later (Bangs and Fritts 1996). Wood Buffalo National Park is therefore the only place in North America where wolves and bison have coexisted in the wild since the Pleistocene.

A second fortune of history is that several frontier naturalists recorded their observations of wolf-bison interactions before both animals were wiped off the plains. These first published accounts of wolf-bison interactions provide a baseline of information about the interactions of wolves and bison prior to European settlement. An important aspect of these accounts is that they describe several fundamental features of wolf-bison interactions that studies of the remnant populations of wolves and bison in WBNP and YNP corroborated more than a century later. One central feature is that bison pose a real threat to the lives of wolves that hunt them. George Catlin, renowned painter and chronicler of Native American culture, witnessed several wolf-bison interactions during his 1832 visit to the upper Missouri River near present-day Pierre, South Dakota. He noted that the "buffalo . . . is a huge and furious animal, and when his retreat is cut off, makes desperate and deadly resistance, contending to the last moment for the right of life—and often times deals death by wholesale, to his canine assailants, which he is tossing into the air or stamping to death under his feet" (Catlin [1844] 2004, 261).

FIGURE 6.1.   The interaction between wolves and bison is the last living example of the megafaunal predator-prey interactions that typified the North American plains during the Pleistocene epoch more than 10,000 years ago. (Photo by Daniel Stahler.)

Catlin's ([1844] 2004, 261) view was shaped by this first-hand observation:

> As one of my hunting companions and myself were re-turning to our encampment with our horses loaded with meat, we discovered at a distance, a huge bull, encir-cled with a gang of white wolves; we rode up as near as we could without driving them away, and being within pistol shot, we had a remarkably good view, where I sat for a few moments and made a sketch in my note-book; after which, we rode up and gave the signal for them to disperse, which they instantly did, withdrawing themselves to the distance of fifty or sixty rods, when we found, to our great surprise, that the animal had made desperate resistance, until his eyes were entirely eaten out of his head—the grizzle of his nose was mostly gone—his tongue was half eaten off, and the skin and flesh of his legs torn almost literally into strings. In this tattered and torn condition, the poor old veteran stood bracing up in the midst of his devourers, who had ceased hostilities for a few minutes, to enjoy a sort of parley, re-covering strength and preparing to resume the attack in a few moments again. In this group, some were reclining, to gain breath, whilst others were sneaking about and licking their chaps [sic] in anxiety for a renewal of the attack; and others, less lucky, had been crushed to death by the feet or the horns of the bull.

Catlin reproduced his sketch as an oil painting—one of the first illustrations of a wolf-bison interaction (fig. 6.2).

Catlin's account may seem exaggerated, but it is con-sistent with contemporary observations of wolves at-tacking solitary bulls in WBNP (Carbyn et al. 1993) and YNP (account 7; fig. 6.3). Both recent accounts depict a violent melee in which the bull fights off wolves with horns and hooves while the predators dart in from the rear to bite his exposed hindquarters. The YNP account accords especially well with Catlin's because it describes the bull striking several wolves and mortally wounding one. Additional reports further demonstrate how bison use physical force to repel and sometimes injure or kill wolves (Carbyn and Trottier 1988; MacNulty et al. 2001; MacNulty 2002).

For bison, physical force is a viable antipredator de-

FIGURE 6.2. A painting titled *Buffalo Hunt, White Wolves Attacking a Buffalo Bull* by frontier artist George Catlin depicting his first-hand observation of a wolf-bison interaction in the upper Missouri River area of present-day South Dakota, circa 1832. The image was originally published in *Catlin's North American Indians* ([1844] 2004).

fense against wolves because they are so much larger than wolves. An adult bison is as much as 10 times heavier, three times longer, and two times taller than an adult wolf (Meagher 1986; MacNulty et al. 2009b). In YNP, for example, bison—which are two to three times larger than elk—stand and confront wolves in nearly 80% of encounters and charge wolves in over 60% of encounters (see, e.g., "Yellowstone Wolves Hunting Bison," filmed by Dan MacNulty in 2007: http://youtu.be/96YxxXEDT-s). By contrast, elk stand and confront wolves in just over 50% of encounters and charge wolves in slightly more than 25% of encounters (MacNulty 2002). Size empowers bison to act aggressively against wolves.

The danger of hunting bison is compounded by their highly gregarious behavior, which includes mutual defense. Catlin (1844 [2004], 260) provided the first record of this: "Whilst the herd is together, the wolves never attack them, as they instantly gather for combined resistance, which they effectively make." Army Captain John C. Fremont (1845, 27) witnessed how a bull bison came to the aid, albeit unsuccessfully, of a bison calf: "There were a few bulls . . . and one of them attacked the Wolves and tried to rescue [the calf]; but was driven off immediately, and the little animal fell an easy prey, half devoured before he was dead." Similarly, Colonel Richard Dodge (1878, 125) recounted his surgeon's observation of six to eight bulls "standing in a close circle" around a calf "while in a concentric circle at some twelve

or fifteen paces distant sat, licking their chaps [*sic*] in impatient expectancy, at least a dozen large grey wolves."

Again, contemporary observations corroborate the historical accounts. For example, in WBNP and YNP bison calves indeed gravitate toward the center of the herd in response to harassing wolves (Carbyn and Trottier 1987, 1988; fig. 6.4). And bulls in both populations do participate in the collective defense of calves (Carbyn and Trottier 1987, 1988; Carbyn et al. 1993; MacNulty et al. 2001). In addition, wolves in YNP are reluctant to attack large bison herds. On average, the odds of attack decrease by 2% with each additional bison in the herd (fig. 6.5). Thus, Catlin's ([1844] 2004) suggestion that herding is an effective defense against wolves is accurate. However, contemporary observations do reveal that the mutual defense behavior of bison has its limits. Specifically, bison provide mutual aid only when it is safe to do so (see hunting accounts).

Because bison usually stand and confront wolves rather than flee, herding allows them to protect their hindquarters from wolves that circle them and attack from behind. Wolves lack the larger size, muscular supinating forelimbs, and longer claws that enable other carnivores such as big cats to quickly grab and overpower prey. Instead, wolves rely solely on their teeth to grab prey and tear into a vital area until the prey weakens and falls. The only location on an adult bison that wolves can easily and safely grab and hold until it weakens is the hind end. Then wolves grab the bison on all sides including nose, neck, and flanks. These are not usually the initial attack points because they are too well protected by the bison's thick hide and fur, sharp horns, and sledgehammer-like front hooves. By herding, bison collectively protect their vulnerable backsides from attack.

However, when alone, a bison standing its ground must turn in place to keep any attacking wolves in sight. This is easy to accomplish when fending off one or two wolves. Larger packs are more difficult to ward off; as a bison turns to charge or kick at wolves that circle it, wolves left unguarded will rush in to grab the hindquarters. If the bison cannot land a blow against one or more wolves to dissuade further attack, it will practically have to spin in place to keep the wolves from biting it. If the wolves can manage to hang on, the bison will eventually exhaust itself. This is one reason why larger packs are more likely than smaller packs to capture bison (MacNulty et al. 2014). Because wolves lack a killing bite, and because a shield of thick hair, hide, and muscle protect bison vital organs, wolves usually start feeding on a captured bison before it actually dies, with an average interval between capture and death of nearly 8 min (MacNulty 2002).

The antipredator benefit of herding for bison is therefore twofold. It dissuades wolves from attacking in the first place, and if they do attack, it neutralizes the ability of wolves to circle and attack from behind. The result is that bison herds are often invulnerable to attack.

Again, Catlin ([1844] 2004, 257) was the first to note this: "While the herd of buffaloes are together, they seem to have little dread of the wolf, and allow them to come in close company with them"

One hundred years later, wildlife biologist Dewey Soper (1941), who authored one of the first scientific studies of bison, including his observations of wolf-bison interactions in WBNP, similarly noted that "bison often exhibit a most curious indifference to wolves" and that they generally "showed no fear" of wolves. Together, large size and herd behavior generally favor bison in the balance of power between wolves and bison. Long ago, Fuller (1960, 16) noted that "bison appear to accept the presence of wolves without panic."

To tip the balance in their favor, wolves must neutralize the advantages of size and herding behavior that bison use against them. Wolves accomplish this by tar-

FIGURE 6.3. Wolves attacking a mature male bison in Yellowstone National Park. (Photo by D. W. Smith.)

FIGURE 6.4. Bison calves move to the center of the herd in response to wolves. (Photo by Daniel Stahler.)

geting small bison (especially calves), solitary bison, and wounded, diseased, or otherwise vulnerable bison (Fuller 1962). Catlin ([1844] 2004) observed wolves to follow behind the bison herds and attack animals that "linger with disease or old age." He wrote that "when the herds are travelling it often happens that an aged or wounded one, lingers at a distance behind, and when fairly out of sight of the herd, is set upon by these voracious hunters" (Catlin [1844] 2004, 260–61). Dodge (1878) and Audubon ([1843] 1897) noted the tendency for wolves to select herds with calves over herds without calves, but it was Carbyn and Trottier (1987) who first quantified this effect: herds with calves comprised only 10% of the herds ranging in WBNP (most herds were bulls only), yet 77% of wolf-bison interactions involved herds with calves.

Wolves also attack bison made vulnerable by terrain that impedes their movement. For example, Audubon ([1843] 1897, 50) noted that "many Buffaloes are killed when they are struggling in the mire on the shores of rivers where they sometimes stick fast, so that the wolves or bears can attack them to advantage; eating out their eyes and devouring the unresisting animals by piecemeal." Similarly, MacNulty et al. (2001) and MacNulty and Smith (2006) described wolves attacking and killing bison moving through deep snow between patches of bare ground made snow free by geothermal and wind action (see accounts below). Bison graze on these patches, but they are vulnerable to wolf predation on the snow-packed trails between patches because the narrow paths and surrounding deep snow hinders their mutual-defense behavior. When bison stray off the trail, they

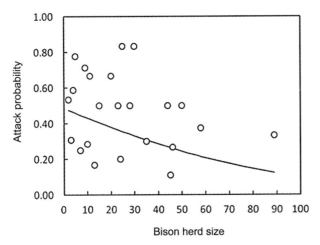

FIGURE 6.5. Effect of bison herd size on the probability that wolves attack when approaching or watching a herd ($b = -0.021 \pm 0.007$, $t = -3.11$, $P = 0.002$; $N = 277$ approaches or watches) in Yellowstone National Park, 1999–2004. Open circles are observed frequencies. Logistic regression was performed on the raw binary data (attack or not) and not the illustrated points, which are provided as a visual aid. (MacNulty and Smith 2006.)

plunge up to their neck in snow, immobilizing them, and making them easier prey for wolves. Deep snow is therefore the ultimate equalizer for wolves; it neutralizes the group defense of bison as well as their individual ability to fight off wolves with their large size (see, e.g., "Yellowstone Wolves Hunting Bison 2," filmed by Dan MacNulty in 2007: http://youtu.be/_yAi8KrvHa4).

Catlin ([1844] 2004, 256) described a similar situation among bison inhabiting the prairies:

> The snow in these regions often lies during the winter, to the depth of three and four feet, being blown away from the tops and sides of the hills in many places, which are left bare for the buffaloes to graze upon, whilst it is drifted in the hollows and ravines to a very great depth, and rendered almost entirely impassable to all these huge animals, which, when closely pursued by their enemies, endeavor to plunge through it, but are soon wedged in and almost unable to move, where they fall an easy prey to the Indian, who runs up lightly upon his snow shoes and drives his lance to their hearts.

Wolves in Yellowstone use exactly the same tactic attacking bison slowed or trapped by deep snow.

The problem for wolves, however, is that they rarely have a chance to attack vulnerable bison. Aside from the fact that young, old, and infirm bison are relatively scarce, those that do exist are usually protected by invulnerable ones (e.g., adults protecting calves [Carbyn and Trottier 1987, 1988; MacNulty et al. 2001]). Moreover, bison try to avoid terrain traps or travel through them quickly where they are unavoidable. The end result is that hunting bison is usually a game of waiting and watching for an opportunity to strike a bison made vulnerable by it straying into dangerous terrain, a lapse in mutual defense, or both (www.press.uchicago.edu /sites/wolves, video 15). Early observers including Catlin ([1844] 2004, 257), Dodge (1878), and Seton (1927) noted that persistent watchfulness was a hallmark of wolves hunting bison. For example, Catlin ([1844] 2004, 257) remarked that wolves "always are seen following about in the vicinity of the herds of buffaloes and stand ready . . . to overtake and devour those that are wounded, which fall easy prey to them." Seton (1927) described wolves hunting bison as "omnipresent" and "tireless."

And because wolves usually cannot kill even a vulnerable bison on their first try, their persistent hunting behavior is also manifested as repeated attacks on the same individual(s) separated by bouts of watching that can last for more than an hour (Carbyn & Trottier 1988; MacNulty 2002). As a result, a single wolf-bison interaction can last days (MacNulty et al. 2001; MacNulty 2002). Catlin ([1844] 2004, 290) was the first to note this: "I have several times come across such a gang of [wolves] surrounding an old or a wounded bull where it would seem, from appearances, that they had been for several days in attendance, and at intervals desperately engaged in the effort to take his life." Wolves in WBNP will stay with herds for up to 6 d before making a kill (Carbyn et al. 1993). That most of this time is spent watching bison, rather than attacking them (Carbyn and Trottier 1987; Carbyn et al. 1993; MacNulty 2002), reflects how wolves are as cautious as they are persistent when hunting bison. Pack size may mitigate the risk of hunting dangerous bison. Carbyn and Trottier (1987) noted that attacks were more likely when packs, rather than single wolves, encountered bison, and MacNulty et al. (2014) found that the probability of attack tended to increase as pack size increased.

When wolves cannot safely attack a vulnerable bison, they resort to scavenging dead bison, which is common. Wolves on the prairies scavenged on bison killed and abandoned by hide hunters (Catlin [1844] 2004) or drowned while attempting major river crossings (Audubon [1843] 1897). In northern YNP, most of the bison eaten by wolves are from scavenged carcasses rather than kills (Metz et al. 2012). And in Yellowstone's Pelican Valley, wolves commonly scavenge bison remains when bison are invulnerable, particularly during mild winters (Yellowstone Wolf Project, unpublished data).

Alternatively, wolves will hunt less formidable prey if it is available. For example, Soper (1941) found moose, deer, and hare in wolf scats in WBNP. Similarly, wolves that usually inhabit Yellowstone's Pelican Valley leave the valley for weeks at a time to hunt elk in northern Yellowstone, where elk remain the preferred prey of local wolves despite nearly equal numbers of elk and bison (Yellowstone Wolf Project, unpublished data).

Hunting Accounts

Published accounts of wolves hunting bison are sparse. Carbyn et al. (1993) observed many such hunts in Wood Buffalo National Park but only published details of two. Smith et al. (2001) described a hunt in Yellowstone. MacNulty has observed 153 hunts in Yellowstone, but because many such hunts can last days, space prevents including all but a few accounts in this chapter. Others are presented in video 16 (www.press .uchicago.edu/sites/wolves) and on YouTube (http:// youtu.be/96YxxXEDT-s [kill no. 1]; http://youtu.be/_ yAi8KrvHa4 [kill no. 2]).

1. February 18, 1980—Wood Buffalo National Park, Alberta, Canada (Carbyn et al. 1993)

At 1310 hr, we located a pack of seven wolves attacking a solitary bull bison. It appeared that the bull had been under attack since 1300 hr and had fought off the wolves at four locations before our arrival. Between 1310 hr and 1330 hr, the wolves attempted to get at the rear of the bull, while he kept spinning around to face and fend off the wolves with his head. The bull ran several times with some wolves hanging on his flanks, rump, and tail. He ran into some willows three times but could only partially shake the intensive attack of the wolves. Attacks were almost continuous, with several wolves darting in from the rear and grabbing the bison's hindquarters, forcing him to turn, at times dragging two or three wolves with him.

The wolves would break from the attack for 15–30-sec intervals, and the bison faced them with his tail up.

Attacks from the rear continued from 1330 hr to 1350 hr. Once we noticed a wolf holding on to the bison's tail for about 10 sec as the bull spun around, dragging the wolf through the willows. The bison appeared to be severely wounded and visibly tiring, although he did not go down. At 1350 hr, all animals were panting heavily; several wolves lay down while the bison stood facing the rest in dense willow. Some wolves were covered with blood around the muzzle and neck, and the bison seemed to be bleeding profusely from the rump and flanks. By 1400 hr, the bull went down.

2. March 24, 1981—Wood Buffalo National Park, Alberta, Canada (Carbyn et al. 1993)

At 0940 hr, the Hornaday River Pack (eight) was running 1 km behind the main portion of a herd of 25 bison, single file for 500 m, led by one wolf at least 100 m ahead. Two groups of bison, one of seven animals, and one of two, were standing in the line of the chase. At 0942 hr, the lead wolf reached the second group of bison, past the seven bulls. The two bison lowered their heads and charged the wolf as it passed them and chased it for 10 m as the wolf continued after the group of 16 still running. Two other wolves passed the two bison (cows) at 0943 hr, but the rest of the pack began to harass these two at 0944 hr. As soon as the first three wolves realized the rest of the pack was attacking the two bison, they ran back and joined the attack. By 0945 hr, all the wolves were circling, lunging, and snapping at the two cows, but one of the cows managed to run from the wolves, whereupon the pack concentrated on the other. Between 0945 hr and 0955 hr, the wolves pressed all around the cow, trying to get at her from the rear, biting the rump and anal area. The cow tried to ward off the attack with her head, but the wolves continued to circle to the rear, so the cow was continually turning in circles.

The cow tried to flee to the herd, which then was at least 1 km away and still moving into open meadows. Trail conditions allowing, the wolves would grab and tear the cow's flanks. Where the trail was narrow, they would follow in file until the trail widened, and several wolves would grab at the cow's hind end. During much of the pursuit, one wolf ran in front of the bison, possibly to distract her or to attempt to grab her muzzle. At 0955 hr, the bison was visibly tiring as she struggled through the snow with several wolves hanging on her hindquarters. The harassment and wounding had taken place over 750 m from the first confrontation to where the cow had dragged herself in among the trees to face the wolves.

By 1000 hr, both wolves and bison were under the trees. The cow had stopped running, but the wolves continued the attack, coming in from behind and wounding her. Two wolves were covered with blood; several were resting as the others continued to press. At 1005 hr, the bison was standing among the trees, her hindquarters shaking. As she fell over once, the wolves closed in immediately and began to tear at her abdomen. The cow managed to get up and dragged herself 10 m with several wolves holding on to her hindquarters. At 1007 hr, the bison was still on her feet, standing and facing the wolves with her tail up. Some wolves were resting nearby, and the bison lunged at several standing close; the wolves stepped back. At 1015 hr, the cow was down, although her head was still up; the wolves were tearing at her abdomen. By 1030 hr, the wolves were feeding. The bison's head rose occasionally, but she was close to death.

3. March 17, 1999—Yellowstone National Park, WY (D. MacNulty and K. Honness)

(To the best of our knowledge, this account is the first record of wolves attacking and killing an adult bison in YNP.)

0724 hr—Thirteen Mollie's Pack wolves approached a bison herd (fig. 6.6) they had harassed the day before. The 58 bulls, cows, and calves occupied a geothermal patch about the size of a football field and watched the oncoming wolves. Nine wolves briefly harassed two bulls at the perimeter but let them go. Wolves lay down next to herd, which remained dispersed and calm.

0744 hr—Two more pack members arrived and bedded alongside other wolves.

0749 hr—Two wolves harassed a cow that stood its ground. One wolf lunged at cow, and was rebuffed by cow's countercharge.

0755 hr—Another wolf harassed the same cow and was also charged away.

0758 hr—Twelve wolves repositioned themselves on a low ridge overlooking herd; three other wolves later joined them. All were bedded. Herd continued to graze, seemingly unalarmed. Seven bulls grazed a windblown ridgetop 50 m from wolves.

1036 hr—One of these bulls walked down ridge toward herd. As he started walking through a patch of deep snow, two wolves sprung from their beds and ran to intercept him. The bull charged, floundering in belly-deep snow. Eleven more wolves joined pursuit. Wolves grabbed and pulled on the bull's rear end, but he bucked them off as soon as he reached bare ground of geothermal patch.

1037 hr—Wolves sat and watched herd which had bunched in response to wolves' provocation.

1055 hr—One wolf harassed four bison at edge of herd. Bison chased wolf back to pack.

1100 hr—Another wolf approached one bull 50 m from the main herd. Bull ignored wolf. Two more wolves approached bull; he stomped front hooves and charged. Wolves retreated, turned, and watched bull. One wolf left bull while two others bedded near it.

1121 hr—Seven wolves approached another bull walking alone toward herd; six wolves charged, but bull escaped into herd. Three wolves approached another outlying bull grazing alone; bull ignored wolves, and they returned to pack. One or two wolves intermittently approached additional outlying bulls with similar results. Meanwhile, the rest of the pack was bedded but watchful.

1224 hr—Herd drifted toward wolves, and one or two leading bulls pushed each wolf from its bed in rapid succession. Wolves moved and lay down outside thermal area. One black wolf circled behind a bull but was quickly run off. Bison grazed where wolves were bedded, having reclaimed thermal area.

1248 hr—Herd started to exit thermal area, walking west away from wolves by way of windblown, snow-free patches interconnected by trails through snow that was variably compacted depending on previous bison movement along route. Bison moved single file along trails to avoid getting stuck in deep snow. A group of bison including 10 calves was first to leave thermal. As they exited onto trail leading to next patch, three wolves rose, stretched, and approached. This startled calves, which hurried to the patch followed closely by adults. Group arrived at patch before wolves could intercept them, and adults encircled calves. Five wolves

FIGURE 6.6. Bison herd on large geothermal area watching arrival of Mollie's Pack in Yellowstone National Park, March 17, 1999. Accounts 3, 6, and 7 refer to this area. In winter, wolves are often reluctant to attack large herds on open patches, preferring to attack them as they move single file through deep snow between patches. (Photo by Kevin Honness.)

sat near group and watched. After brief rest, two wolves harassed one cow and two calves standing at edge of herd. Cow charged, and calves stood behind cow. Wolves retreated and resumed watching. Group tightly bunched.

1259 hr—Group filed onto trail leading 150 m west to next patch. Eight wolves rushed group, which stampeded along trail. Calves were in front, adults at rear. Wolves stopped as group reached patch; they abandoned group and returned to bison still on thermal area.

1304 hr—Seven wolves harassed one bull, which ran and joined six others. Wolves let him go and harassed two other bulls. One fled and joined remaining cow-calf group of 10, and one stood and repeatedly charged until wolves relented.

1319 hr—Cow-calf group, which included some bulls, exited thermal, heading west. Group ran single-file on same trail traversed by first group at 1248 hr. Nine wolves rushed, dodging adults at rear that charged

them. Group reached patch and bunched up. As this interaction transpired, the first group fled farther west, hurrying along a trail to the first of a series of snow-free bluffs.

1321 hr—All 15 wolves caught sight of first group and headed after them. Some ran, some walked, producing a haphazard procession across the snowy plain. Bison reached bluff before pack could catch up. Eighteen bulls, cows, and calves stopped and bunched as others fled to next bluff. Five wolves reached the 18 and swarmed them. Adults encircled calves and thwarted wolves.

1335 hr—As the 18 walked slowly from wolves, two cows and one calf fell behind. Wolves rushed bison and lunged at calf. Cows defended calf and all three fled back to group with wolves in pursuit. Wolves stopped once bison joined herd.

1336 hr—Remainder of pack arrived, and all 15 rallied. Afterward, they lay down and watched bison group gathered at edge of bluff where trail led to next snow-free bluff to the west.

1344 hr—Thirteen bison, including two calves, started on trail and rushed single file to next bluff. Thirteen wolves ran after them and grabbed hind end and flanks of last bison (cow). She pulled them 50 m to the bluff, where two bison came to her aid and drove off attackers.

1346 hr—A second line of bison, including 12 adults and three calves, came charging on trail toward bluff. Thirteen wolves circled behind line and grabbed trailing bison, another cow. They caught her where they grabbed the cow at 1344 hr, and like that one, she dragged them to the bluff, where she shook them loose. But her backside was torn and damp with blood.

1348 hr—Twenty-five bison, including the wounded cow, raced across bluff and onto trail leading to next snow-free bluff to the west. Thirteen wolves chased the bison and 10 of them grabbed another cow; one grabbed nose and nine grabbed hindquarters. She dragged them 15 m to the bluff, where the herd now stood. One bull ran out and charged off the wolves. She rejoined the herd.

1350 hr—Wolves left herd and returned east to a group of seven bison (six bulls, one cow) remaining on wind-scoured bluff described at 1346 hr. Four wolves circled cow and lunged at her hindquarters. One wolf lunged at her nose but missed. She turned, kicked, and charged. Two bulls also charged. Wolves stopped and watched bison (fig. 6.7A).

1357 hr—Cow and two bulls started walking west, away from wolves. As they did, 11 wolves simultaneously rose up and approached them. The two bulls left cow; one walked east, past the wolves joining the other four bulls, and one walked west toward bluff's edge. Cow was swarmed by wolves (fig. 6.7B). She repeatedly pivoted on her front hooves, swinging her rear end 180° and kicking out with hind hooves. Wolves could not grab her. As she spun and wheeled, she inched toward bull that had walked west to bluff's edge. He stood and watched her acrobatics until she was so close that wolves harassed him. He joined the fray, kicking and charging. One wolf dodged bull, lunged, and grabbed cow's rear, but she kicked fiercely, and wolf went flying. Wolves relented and stood scattered around cow and bull.

1401 hr—Cow turned from wolves and walked west. As she stepped on trail, wolves swarmed her. She pulled back and fled east toward five bulls. Bull that had accompanied her walked west. Wolves momentarily flocked to bull, then reversed course when they saw cow fleeing east. She was alone and wheeled about as before, trying to keep 10 swarming wolves at bay. She hooked one wolf with a horn and nearly threw him. One wolf bit her rear. Panicked, she sprinted to five bulls, which gathered around her and chased wolves away.

1403 hr—Wolves returned west to bull, which grazed bluff edge. Another bull grazed nearby. Wolves lay down around bison and watched them. A third bull came into view from behind bluff. Twelve wolves harassed this bull, which joined other two. Together they kept wolves at bay. Wolves relented, and two bulls walked east, joining the five bulls and one cow.

1417 hr—Herd of 25 mentioned at 1348 hr walked west on snow-packed trail toward a snow-free bluff. Wolves saw this and approached.

1428 hr—Herd reached bluff, and wolves returned east to bulls and cow, which grazed open ground at top of bluff. Group included eight bulls, one cow. Wolves lay on snow beside them.

1434 hr—Seven wolves harassed bison as others watched. Group stood together against wolves. One bull charged three times. Wolves withdrew and bedded nearby. One wolf rose, briefly harassed one bull, and retreated. Bison drifted west, away from wolves.

1447 hr—Three wolves harassed cow and one bull. Both lingered behind as other bison walked west. Bull charged and dispersed wolves. Unclear which bull.

1449 hr—Herd of 25 descended snow-free bluff and started west on trail. Wolves saw this and approached herd but turned back to cow and bull when herd turned back and reoccupied bluff.

1456 hr—Two wolves harassed the cow and bull. Bull walked west and left cow. Ten more wolves converged on her. She sprinted to bull, and they stood off wolves.

1500 hr—Wolves withdrew and rallied, jumping on and tackling one another. Eight wolves then resumed harassing cow; bull stood beside her. Wolves circled and lunged at cow; she charged and kicked at them. Bull mainly stood still. Wolves withdrew and rested.

FIGURE 6.7. Wolves resting and watching bison between attacks (*A*) and attacking an isolated adult female (*B*) in Yellowstone National Park, March 17, 1999. See account 3 for details. (Photos by Kevin Honness.)

Wolves alternated between harassing cow and resting for 40 min.

1540 hr—As wolves rested, cow and bull ran to next bluff and joined with one bull grazing there. Most wolves seemed asleep because only three wolves pursued.

Meanwhile, the herd of 25 vacated bluff occupied since 1449 hr and hurried single file west on trail.

1551 hr—One wolf saw herd, pursued, and caught up to it as the last bison in line, a cow, was nearing the snow-free top of a steep bluff. Wolf grabbed cow's tail, and she pulled him into the patch where other bison stood. They chased away wolf.

1554 hr—Five more wolves trotted west toward herd. They intercepted seven bison, including one calf that lagged on trail behind herd. Bison grouped tightly; one bull charged.

1558 hr—Rest of pack arrived; 15 wolves lay around seven bison.

1612 hr—Herd departed steep bluff and ran single file on a trail that led 160 m west to next patch of bare ground. Bison moved slowly on trail because it was not well packed. One wolf gained on last bison, a small adult (probably a cow), and grabbed its hind end. Four more wolves caught up. Together, five wolves pulled on rear of bison for 1.5 min as it dragged them to next patch. A raven, anticipating its next meal, hovered overhead. But bison reached patch, turned, and scattered its attackers. Wolves rallied, greeting each other excitedly. Afterward, they harassed the herd but were rebuffed.

1631 hr—Group of seven bison from 1554 hr ran single file toward herd and wolves on same trail that herd traversed at 1612 hr. Wolves rushed the group, circled behind, and grabbed the rear and flanks of the last bison, a cow. Wolves passed one calf in center of line. Cow pulled herself to patch and rejoined herd in <2 min. One bull charged and scattered wolves east. About 32 bison stood wedged together on tiny island of bare ground in a sea of deep snow. Trail to the next patch was 330 m, the longest yet.

1633 hr—Wolves mingled 20 m east of patch, seemingly undecided about their next move. Nineteen bison (bulls, cows, and calves) took advantage of the pause and fled on 330-m trail heading west to next patch. The bison sank chest deep in soft-packed snow on trail. Wolves immediately noticed fleeing group and ran after them. From our vantage 2 km away, the line of bison resembled a train of boxcars pushing straight across a frozen prairie. In <1 min and about 50 m down trail, 11 wolves grabbed the hind end of "caboose" cow. She tried to pivot 180° to protect her hindquarters, but trail too narrow to maneuver. Wolves overwhelmed her. She regained trail and marched after the herd, which was distancing itself from the cow. One wolf released cow and sprinted after herd. A hard snow crust buoyed wolf as it ran ~100 m and grabbed hind leg of second-to-last bison, a calf. A bull was last. Ten wolves still bit cow, tearing away hide and hair, but she dragged them behind her.

1635 hr—As wolf fought to topple calf, bull behind it walked past, ignoring calf's predicament. One wolf left cow, ran to calf and grabbed it alongside the other wolf. Calf inched toward end of trail where herd was waiting on patch of bare ground.

1638 hr—Calf reached patch, and two wolves released it. Other bison stood watching. One wolf returned to cow, still dragging 5–6 wolves. Other wolves scattered on trail behind cow, eating blood-soaked snow and other scraps left in cow's wake.

1639 hr—As cow approached to within <10 m of patch where herd stood, wolves peeled away and released her; they were preoccupied with eating scraps on trail. They had seized her for nearly 5 min, and their damage was clear. Large crimson wounds flashed brightly on cow's rear and flanks in afternoon sun. Her gait was stiff as she walked, then she disappeared into herd.

1653 hr—Herd fled patch, heading west on trail toward next patch, about 120 m away. Wolves first chased herd, then pivoted toward patch where herd had been. There lay the wounded cow, and standing beside her was one bull. Wolves swarmed cow's backside and started feeding, 9.5 hr after they first approached the herd at the thermal, 2.5 km away. Bull stood watching; wolves ignored him. Herd continued west, away from wolves.

4. March 24, 2000—Yellowstone National Park, Wyoming (D. MacNulty and N. Varley)

The account that follows (MacNulty et al. 2001) describes the final attack of a wolf-bison encounter that began March 23, 2000. Five Mollie's Pack wolves encountered a herd of 59 bison on March 23 at 0908 hr, and a grizzly bear appeared at 1231 hr, running toward the wolves as they chased a bison. In each of two subsequent attacks observed that day, the grizzly ran immediately behind or alongside pursuing wolves. Two additional adult-sized grizzlies, traveling together, approached the wolves at 1838 hr. Subsequent play behavior suggested that the bears were a 2-yr-old sibling pair.

On March 24, 1213–1819 hr, one to five wolves made a series of 14 attacks, alternately attacking and resting, on a bison group containing nine adult/juvenile bulls and one 11-mo-old calf of unknown sex. Group remained in the area after herd fled when pack pursued one unidentified wolf approaching bison at 1119 hr. Unidentified wolf was not wearing a radio collar and was presumed to be either an unrelated trespasser or an unwelcome relative. Pack mobbed wolf but allowed it to escape.

Wolf attacks on bison lasted 2–11 min and involved wolves lunging at bison while they stood grouped on a thermal area or running single file through deep snow between thermal areas. Wolves targeted calf during all attacks. In four attacks, one or two wolves grabbed calf and inflicted visible damage to its rump and flanks but failed to kill it.

In all but the final attack, one to five bulls defended calf by charging and kicking wolves.

The final attack was preceded by a 3.5-hr standoff between wolves and bison on a thermal area at the foot of a small snow-covered ridge. During this period, wolves made periodic attempts to grab calf but were repelled each time by bulls. Standoff ended when seven bulls walked away from thermal area and ascended ridge, leaving two bulls to defend calf. After another failed attempt to grab calf, wolves left thermal area at 1805 hr and trotted toward seven bulls. Meanwhile, one grizzly appeared and approached two bulls and a calf.

> 1814 hr—Two bulls, followed closely by calf, left thermal area and ascended ridge single file toward seven bulls and five resting wolves. Grizzly continued approaching.
>
> 1815 hr—Sighting oncoming bison, pack arose and approached two bulls and calf from the west while the grizzly approached from the east. Calf struggled through snow and trailed 10–15 m behind two bulls.
>
> 1819 hr—Five wolves walked past two bulls, and four wolves grabbed calf; three grabbed its hind end, and one grabbed its flank. Seconds later, a fifth wolf grabbed its neck. Two bulls continued walking and joined seven bulls.
>
> 1820 hr—Grizzly continued to approach from the east, nearing calf and wolves.
>
> 1821 hr—Three wolves released calf and charged the grizzly while the other two still grabbed hind end of calf.

When wolves lunged at the grizzly, it rose on its hind legs, quickly fell back to all four feet, and fled with three wolves in pursuit. Within seconds, the grizzly turned, ran past pursuing wolves, and headed back to calf.

> 1822 hr—Grizzly forced two wolves from calf's hind end and attacked calf with its forepaws. Three wolves grabbed front end of calf.
>
> 1823 hr—Three wolves bit front end of calf while grizzly swatted at its hind end.
>
> 1824 hr—Grizzly pulled calf from the wolves, dragged it to the ground, and began to feed. Wolves stopped contesting calf. Four wolves disappeared over ridge while one wolf stood by and watched the grizzly feed.

On March 26, the wolves abandoned the kill to the bears and continued hunting bison as follows.

5. March 26, 2000—Yellowstone National Park, Wyoming (D. MacNulty and N. Varley)

> 1147 hr—Five Mollie's Pack wolves approached a herd of 35 bison (mainly cows and calves). Herd was 300 m west of where cow was killed in account 3.
>
> 1154 hr—One wolf approached to <10 m of one cow; other wolves bedded.
>
> 1155 hr—Two more wolves approached cow.
>
> 1156 hr—One wolf harassed cow at edge of herd.
>
> 1158 hr—Second wolf joined and harassed cow; she charged twice.
>
> 1159 hr—Four wolves rallied. Cow shook its head and charged.
>
> 1200 hr—Wolves approached calf; cows encircled it.
>
> 1201 hr—Four cows charged wolves.
>
> 1202 hr—Wolves bedded.
>
> 1209 hr—Two wolves harassed two cows on edge of herd. Bison approached wolves and drove them off.
>
> 1223 hr—Pack abandoned herd.
>
> 1240 hr—Pack approached a second herd (19).
>
> 1244 hr—One wolf approached herd and was chased away. Other wolves bedded and watched.
>
> 1257 hr—Three wolves approached herd. Two wolves lay beside it and watched herd graze. The third wolf circled behind herd.
>
> 1306 hr—Three wolves abandoned herd and approached small herd of juvenile bulls on nearby windblown ridgetop.

1310 hr—Two wolves harassed four juvenile bulls atop windblown ridge.

1311—Bison charged. One wolf persisted—jumped in, jumped out, drew mature bulls out. Second wolf circled behind bison. Wolves isolated two juvenile bulls. A third wolf arrived. One bull kicked and struck this wolf. Two other juvenile bulls stood to the side.

1312 hr—Three juvenile bulls panicked, ran single file down windblown ridge, and plunged into deep snow. The three wolves rushed lead bull and bit his hind end. Bull immobilized in deep snow. Two more wolves arrived and bit the hindquarters of second bull. Third bull managed to turn around and returned to herd on ridge. Second bull also turned around—despite two wolves clinging to its hindquarters—pulled itself and two wolves back up ridge to herd. The two wolves released bull and joined other three wolves in attacking the first bull, still stuck in snow.

1318 hr—First bull still fighting for his life. Five wolves bit and tore at it.

1341 hr—Wolves fed on young bull, which appeared dead. Wolves filled up before one grizzly bear usurped the carcass at 1921 hr.

6. March 30, 2002—Yellowstone National Park, Wyoming (D. MacNulty)

0545 hr—Ten Mollie's Pack wolves rested 50 m from 53 cow, calf, yearling, and 2–3-yr-old bull bison. Pack included six radioed individuals—two adult males (193M, 194M), two adult females (174F, 175F), two male pups (261M, 262M)—plus four other pups and yearlings. Bison lay compactly on an outermost patch of snow-free ground above large geothermal area, the same one described in account 3 at 0724 hr (fig. 6.6).

0600 hr—Wolves left bison herd but returned at 0715 hr.

0718 hr—Wolf 261M casually lunged at a few bison, which stood together and shook their heads laterally, a sign of agitation. Wolf 261M retreated and bedded with pack, closely watching herd.

0725–30 hr—Three cows and one juvenile bull walked single file to nearby snow-free and bison-free ground 40 m away, grazed, and were joined by three cows. Wolves lay scattered between the two patches.

0732–34 hr—Eleven more bison filed from first patch to second. As calf entered snow, Wolf 194M sprung up and charged calf, which fled back to patch, with 194M close behind. A cow there rushed to calf and drove wolf away. Meanwhile, the nine other wolves rushed the 10 bison still walking toward second patch. Bison stampeded back to first patch. Wolves 193M and 194M plus one other grabbed a cow by the rump but released it as she reached herd on patch.

0734 hr—Wolves sat or lay watching. The 40 bison stood on their snow-free patch, their only escape route through the wolves. A few bison grazed on sparse stubble.

0813–46 hr—Bison relaxed, spread out, and grazed along a windblown slope extending downhill opposite wolves. Sixteen bison inched off to graze in smaller patch about 10 m away.

0850 hr—While bison grazed farther downslope, the pack entered first patch, stood, and watched them crowding opposite edge of patch, some grazing in shallow snow just beyond.

0913–26 hr—Wolves 261M and 262M and two others approached the 16 bison grazing in a smaller patch. They continued grazing. A cow approached edge of patch, stopped, and stared at 261M, with her tail raised in agitation. Other bison grazed on while the four wolves stood and watched. Wolf 261M crossed into smaller patch, and a cow immediately chased him to the snow. Wolf 261M turned toward cow, the two standing nearly nose to nose. Meanwhile, other bison in the smaller patch rejoined main herd on first patch. Wolf 261M and another wolf tried to enter patch again, but cow charged them away. Remainder of pack rested near main herd, apparently uninterested in 261M's activity.

0926–1133 hr—Bison grazed; wolves watched and socialized; a few approached seven bison grazing on smaller patch but no escalation.

1133 hr—Most bison bedded compactly on bare slope; 14 stood; five grazed; three grazed in smaller patch; juvenile bull grazed alone in snow beyond smaller patch.

1143 hr—Wolves arose and rallied. Bison mostly bedded in one group, except juvenile bull, still grazing alone.

1145 hr—Pack approached herd, and four wolves approached juvenile bull, blocking its escape to herd.

1150–57 hr—Juvenile bull raised its head and tail and fled through wolves toward herd but stumbled in snow, and the pack swarmed him. All but two grabbed him; 175F and pup hovered nearby and watched. Bull struggled through snow with nine wolves latched to flanks and rear. As he neared herd, a cow charged wolves and was backed by 20 others from herd walking shoulder to shoulder through snow. As these bison neared the melee, three or four wolves, caught between bull and its rescuers, released bull and fell behind the five or six wolves still biting him. But bull suddenly escaped to herd, and lead cow lunged with front hooves, scattering pack and nearly trampling 174F and 261M.

1157–59 hr—Wolves moved from herd and rallied. Meanwhile, nine bison sniffed the bloody snow where the bull had been attacked. Herd bunched on patch around the standing, wounded bull.

1159–1220 hr—Wolves lay down on bison's escape route, a trail through snow leading to large geothermal area at base of hill. This bare, open expanse provided safety because it allowed bison to aggregate and confront wolves.

1220–25 hr—Wolf 262M slowly approached within 5 m of three juvenile bulls grazing on separate patch. One bull, shaking its head and raising its tail, at edge of patch lunged at 262M with front hooves, but wolf remained safely in snow, where bull dared not tread. Each time bull grazed, 262M pressed forward, and bull charged.

1225 hr—Wolf 261M plus a pup noticed 262M's action and approached three juvenile bulls. Two fled toward herd. The three wolves surrounded last (largest) bull, which kicked with its hind hooves, then also fled to herd, still standing bunched on windblown slope. Wolf 261M grabbed bull's hind leg but lost hold.

1226 hr—Wolves continued watching and waiting. The herd bunched but was restless.

1227–30 hr—Juvenile bull crested ridge above herd. Wolves 261M and 262M and a pup immediately rushed him, and 261M tried to grab his hindquarter. Wolves 193M and 194M joined attack, and one also grabbed his hindquarter. Bull kicked wolves off and escaped into herd.

1230 hr—Wolves retreated and sat.

1231 hr—Mature bull that had been grazing in patch on opposite end of trail that wolves were guarding started along trail toward herd; six wolves (194M, 261M, 262M, and two noncollared black pups) im-

mediately surrounded him. He raised his tail, and as four more wolves arrived, lashed out with his front hooves, driving wolves back. Whenever a wolf came close to bull's hindquarters, he snapped a hind hoof backward with tremendous force. Wolves withdrew beyond bull's range and surrounded him. Bull bluffcharged, and all wolves flinched and retreated farther.

1234 hr—Although bull stood in snow, wolves relinquished the attack and lay down. Bull resumed walking toward herd. Three wolves followed; bull stopped, looked at wolves, and wolves lay down. Bull started again, and all 10 wolves surrounded him. Bull ran, and wolves pursued; bull suddenly turned and rushed wolves. Wolves dodged bull; he returned to trail, slowly walking toward herd, with wolves fanned out behind. Bull and pack stood still. One wolf lay in trail directly in front of bull. Bull slowly turned his head left and right, toward each wolf around him. He walked on, and wolves followed. Bull stopped, looked at wolves, and they stopped.

Bull walked on, and wolves rushed in behind. Two jumped in front of bull, and he hopped forward. Snow deepened, and bull sank to his belly, while wolves stayed on top, biting bull's flanks and hindquarters. Bull kicked out with rear hooves, knocking wolves backward or driving them into snow. For each wolf that fell, another took its place; pack covered him like a cloak. As bull pushed forward, snowpack thinned, and he regained footing, confronted pack, and kicked back at wolves circling behind. One wolf dodged a kick by ducking beneath bull's outstretched leg. Bull alternated between fleeing toward herd and confronting wolves.

1240 hr—As bull reached bare patch with herd, three cows and one juvenile bull approached. Wolves retreated, rallied briefly, bedded on snow and watched bison graze. Four more bison joined first four.

1243 hr—Bison grazing at edge of patch nearest wolves suddenly pulled up, initiating group's retreat to herd, with 10 wolves pursuing. The eight bison reached the herd before wolves could grab them. Wolves withdrew from patch, bedded in snow, and resumed watching herd, which remained trapped on windswept slope.

1246 hr—Pack walked a short distance and harassed a second mature bull grazing on patch alone. Bull charged wolves, which retreated and bedded down around bull. Bull resumed grazing with pack scattered around him.

1307 hr—Herd relaxed; individuals grazed windswept slope, and many left patch, crossing a 10-m snow band, and entering smaller patch at toe of slope, beyond which was only unbroken snowfield. Some moved to edge of snowfield to graze where wolves first grabbed juvenile bull at 1150 hr.

1308 hr—Entire pack trotted downhill toward bison at edge of snowfield, pausing 25 m away, then lay on their haunches and stared at bison. Bison continued grazing, apparently unaware of wolves. Wolves 193M and 262M and another wolf walked onto windswept slope, lay at very top of patch, and watched herd below.

1315 hr—Bison at snowfield edge detected wolves and fled uphill toward windswept slope. All bison rushed backed to windswept slope, emptying small outlying patch. Bison pushed wolves off patch, and most walked a short ways in snow and bedded. One wolf lay <10 m of herd and watched it. Bison were jostling each other, with larger animals pushing smaller ones to side.

1318 hr—Five wolves left herd and headed to large patch at base of slope, where a single, mature bull grazed. But wolves rested.

1322 hr—All wolves gathered on large geothermal patch at base of hill.

1336 hr—Herd drifted back down slope and toward small patch near edge of snowfield, which they had vacated at 1315 hr as wolves approached. Herd pushed past outlying patch into snowfield toward windswept ridge on opposite side of drainage. Snow was belly deep on mature cows, and distance to the ridge at least 500 m. Within minutes, bison fled back to windblown slope. Wolves apparently were unaware of herd's attempted escape.

1346 hr—Pack slowly walked back uphill and bedded in snow near herd, blocking its escape route. Herd packed together at top of windswept slope.

1404 hr—At least nine wolves investigated bison trail into snowfield where bison had attempted escape at 1336 hr. Herd started leaving slope, and one wolf charged up slope toward herd. Herd ran through snow farther uphill into a more distant bare patch.

1406 hr—Pack watched herd graze at new, much larger, bare patch. Herd continued uphill farther from wolves.

1409 hr—Pack charged herd, which stampeded uphill along broad, open ridgetop. One wolf attempted to grab a calf trailing the herd. Five more converged on calf. Calf kicked back at wolves, which dodged the blows; calf escaped into herd that had faced wolves. Snow was packed, providing bison ample mobility. One cow rushed out at wolves, which scattered in different directions. Entire herd pushed toward wolves, which backed away. Wolves sat in snow and watched herd graze on patch. Wolves <10 m from nearest bison. Bison nervous, pushing and head butting.

1423 hr—Two wolves approached herd spread along slope grazing. Remaining wolves joined lead wolves. As wolves approached, bison at rear of herd fled uphill toward the part of the herd grazing farther uphill. As bison ran, all wolves pursued, fanning out behind. Herd reached highest point of open slope, where two mature bulls were grazing. Herd confronted wolves, which halted and watched as herd began to graze.

1425 hr—Mature bull drifted downhill, grazing as it walked. Two wolves, including an adult male, moved into snow out of bull's way. Five bison from herd followed bull through wolves. Many others followed bull. Three more wolves moved out of bull's way. Herd slowly pushed down the broad, open slope, almost imperceptibly, as wolves stood and watched. A spacious patch granted bison maximum freedom to maneuver.

1426 hr—Cow at rear of herd suddenly ran downhill, triggering entire herd to do likewise. Pack stood and watched. Herd slowed as it passed them. Three bulls remaining upslope grazed as wolves looked on.

1430 hr—Herd started crossing snow toward windswept slope. Ten wolves ran to herd, which stampeded downhill to trail and then to large bare patch at base of hill, where wolves had attacked mature bull at 1234 hr. Deep snow slowed bison, which ran single file through snow. Larger bison pushed onto trail, knocking smaller ones into snow. Other bison waited to get in line; seven bulls turned back to patch. Wolves passed them while chasing main herd.

1431 hr—As wolves neared, five or six bison at rear of herd left trail and ran abreast, plunging through snow with wolves at their heels. Meanwhile, herd reached bare ground and ran downhill through patch and toward trail leading through snow into next patch. Calf fell on open slope, recovered, but dropped back to rear of herd with wolves closing. As bison reached the next patch, 13 ran onto trail out of the patch, and 31 turned back. Three wolves chased the first group, which struggled through belly-deep snow, while other seven wolves harassed second herd bunched on patch.

1432 hr—As the first group descended into a gully and disappeared, the second group ran out of patch and onto trail with seven wolves pursuing. Bison spilled across trail, several abreast through snow. Many tripped and tried frantically to recover. A cow leapt over a juvenile bull when he stumbled. When bull emerged he trailed herd, with wolves bearing down. He sprinted ahead of wolves and wedged himself between two other trailing bison just as chase led out of view into gully.

1433 hr—Both groups emerged from gully and ran to geothermal patch at base of hill.

1436 hr—A juvenile bull, perhaps that noted at 1432 hr, emerged from gully pulling 10 wolves gripping his rump, flanks, and neck. He shook off four or five wolves and reached the trail through a snowfield. He slowed and started to stumble. Wolves pulled furiously on him. He fell forward, was momentarily prone, then lurched forward and continued on with six wolves tugging on him and three others jumping back and forth on trail before him. Another wolf lay in trail behind, eating bloody snow and flesh that the pack had pulled off. Bull's circumstances worsened as trail through snowfield deepened, forcing him to negotiate deep foot wells left by herd's passing. Herd grazed peacefully about 30 m away on large geothermal patch below.

1440 hr—Bull faltered and fell into trail leading to herd. After a few seconds, he arose and struggled forward with wolves clinging to his backside. One wolf tugged at his nose.

1444 hr—Bull repeatedly rested and struggled forward as wolves bit. They released him for a moment, before grabbing him again. Bull fell a final time, with all 10 wolves biting him.

7. March 6, 2003—Yellowstone National Park, Wyoming (D. MacNulty)

0706 hr—Four of eight Mollie's Pack wolves approached a solitary bull bison standing on a small patch of bare ground on slope above large geothermal area. A herd of 45 (bulls, cows, and calves) was bunched on the geothermal. Bull looked as if it had been attacked recently. Its tail was crooked and its hind end was misshapen, suggesting wolves had pulled and bitten both. Five to 10 ravens fluttered about bull.

0706 hr—One wolf circled behind bull; he stood and slowly turned his enormous head to track wolf behind him. All four wolves drew closer to bull.

0708 hr—Surrounded, bull charged, pawed ground, and kicked. Wolves lunged, retreated, and lunged again, akin to a fencer's lunge and parry.

0710 hr—Wolves rallied, greeting each other. Adult female Wolf 174F appeared with a severe limp. Her right foreleg was swollen to about twice the width of her left foreleg. Previous observations indicated she sustained this injury about 2 wk earlier, probably when hunting bison.

0711 hr—Seven of eight wolves harassed bull. Wolf 174F was involved despite her injury. Bull stood his ground, charging, kicking, and pivoting to keep wolves at bay. His movement was stiff and slow, maybe due to an earlier, unobserved attack. Nevertheless, he delivered many potentially lethal blows. He kicked several wolves aside and hooked one with his horns and sent it airborne. But for each wolf he kicked or charged, another would lunge from behind and grab his backside. He bucked and kicked them off. Wolf 174F grabbed his rear but was also kicked off.

0714 hr—Wolves stopped and watched bull; they stood or lay beside him. One scrambled away when bull charged it. Bull pawed at packed snow beneath his feet and nibbled a few meager grass stems. Twenty ravens gathered around bull, irritating him; he occasionally charged and kicked them, too.

0748 hr—Three wolves lunged at bull. Bull kicked but made no contact.

0749 hr—Wolves retreated and then rallied.

0750 hr—Four wolves returned to bull, briefly harassed him, lay down, then rose and harassed him again. Bull stood and kicked at wolves.

0754 hr—Wolves retreated, rallied, and resumed watching bull.

0802 hr—Two wolves lunged at bull; he charged them away. The two rallied briefly; afterward they settled down near other wolves, and all watched bull. Wolves alternated between attacking and watching bull for next 1 hr and 25 min. One to six wolves attacked eight times. Bull stood and defended himself each time.

0927 hr—Pack left bull and approached herd of 45 bison that grazed on geothermal area at base of slope. Wolf 175F (sister to 174F) was limping. She did not step with her right rear leg but kept it tucked under her body as she walked. Given that she was not limping earlier we surmised that she was injured during melee with bull.

0951 hr—Wolves approached bull grazing at edge of herd. He turned and walked into herd. Six wolves attacked another bull at herd edge. He pivoted, charged, and drove wolves away. Wolf 175F was bedded nearby, not participating.

0955 hr—Wolves returned to solitary bull; he had not moved from where wolves left him at 0927.

1000 hr—Seven wolves attacked bull, including 174F. Wolf 175F lay down nearby. Wolf 174F mainly stood in front of bull, seemingly baiting him to charge, as others lunged at his backside.

1011 hr—Attack stopped; wolves rested and watched bull.

1031 hr—Wolves drifted east of bull, some lying, others playing. Bull lay down.

1043 hr—Six wolves approached and harassed bull; 174F and 175F still bedded. Bull stood up and turned to keep wolves to his front side.

1045 hr—Wolves resumed watching bull.

1057 hr—Wolves walked back down to herd. Wolf 175F limped part way and then bedded. Bull also lay down. Herd was grouped but not tightly so. Five cows walked out from herd and met first wolf, which circled around herd and inspected a bull grazing 30 m from herd. Herd started to pack closer together. A lead cow charged one wolf. Wolves stood and watched. Five wolves approached outlying bull. He shook his head laterally and charged one wolf. Two wolves lay near him, and he ran them off. Six wolves chased five cows back to herd.

1140 hr—Wolves abandoned herd and meandered east. Solitary bull remained bedded on slope above. Snowfall obscured our view, but it seemed wolves were drifting back to bull.

1415 hr—Weather cleared, revealing four wolves surrounding bull. He stood facing them, his backside clearly torn and swollen. Six wolves, including 174F, attacked. Bull defended himself as before.

1418 hr—Wolves retreated, rallied, and lay around bull. He was caked in fresh snow except for his rear end, where warm, leaky wounds kept snow from accumulating.

1436 hr—Bull lay down, apparently exhausted. Within seconds, three wolves rose and approached. Bull stood up and confronted wolves, which stood back and resumed watching.

1437 hr—Bull lay down; one wolf rushed, and bull stood back up. Wolf lay in front of bull. Bull's tail hung limply.

1454 hr—Wolves walked back to herd and approached an outlying bull, which charged them away. Wolves meandered around herd for about 20 min, then started back toward bull.

1523 hr—Through heavy snowfall we saw vague outlines of bull stumbling, falling down, and rising to his feet. At least one or two wolves grabbed his backside. He dragged them as he wheeled about.

1527 hr—Snowfall completely obscured our view.

1557 hr—Between snow squalls we could see that bull was still standing and keeping wolves at bay.

1722 hr—Six wolves bedded around bull. Each was lying <10 m from bull. The snow around him was peppered in blood, most likely his own.

1728 hr—As bull lay down, one wolf rose up and rushed bull. It started chewing on the bull's rear end. Bull immediately stood up, chased wolf away. Other wolves were bedded or standing around bull.

1830 hr—Five wolves harassed bull; he repelled them.

1838 hr—Wolves lay around bull, watching him. One wolf sneaked behind bull and briefly grabbed his backside. Four more wolves sprang up and started lunging at bull. One wolf dashed back and forth in front of bull, and others crowded behind him.

1841 hr—Wolves stopped attacking and rested beside bull.

1844 hr—Six wolves rallied, and afterward they surrounded bull. Bison stood, watching them until darkness at 1849.

Next morning (March 27) at 0645, wolves were feeding on bull. Wolf 175F was found dead several days later,
curled up beneath a large spruce tree 2 km from bull
carcass, apparently having succumbed to injuries from
the bull.

Conclusion

The preceding accounts demonstrate that killing bison
is near the very limit of wolf predatory power. The success rate of wolves hunting bison (3%–10%) is the lowest
reported for any of their prey (Carbyn et al. 1993; Smith
et al. 2000; MacNulty 2002; Mech and Boitani 2003).
The success of wolves hunting bison in YNP (7% of
74 wolf-bison encounters) was somewhat less than in
WBNP (10% of 31 wolf-bison encounters), but the difference was not statistically significant (MacNulty 2002).
This is noteworthy considering that wolves reintroduced
to YNP had little to no previous experience hunting bison
(Smith et al. 2000), while wolves in WBNP have always
killed bison (Carbyn et al. 1993). Lack of significant difference in hunting success between the two populations
suggests that wolves rapidly learn how to hunt novel
prey. This is not surprising given that the potential price
of failure when hunting large dangerous prey is death.

An interesting and provocative aspect of the historical accounts of wolf-bison interactions is the purported
large size of packs. According to Catlin ([1844] 2004],
261), wolves "often gather to the number of fifty or
more." Similarly, Fremont (1845) observed a bison calf
pursued by "20 or 30 wolves." Although Catlin's observation seems outrageous, contemporary studies indicate
that pack size with ≥20 wolves is not unheard of. From
1995 to 2011, seven of 38 wolf packs in YNP reached or exceeded 20 wolves, including one pack of 37 wolves. Two
of the seven frequently preyed on bison. Notably, five of
seven of these instances of large packs occurred prior to
2005 when the Northern Range elk herd exceeded 9,000
animals. In the case of plains wolves and bison, the notion that large packs were common is plausible given
that "the bison population west of the Mississippi River
at the close of the Civil War numbered in the millions,

probably in the tens of millions" (Shaw 1995). The largest
pack ever recorded (42 wolves) inhabited Wood Buffalo
National Park (Fau and Tempany 1976, cited in Carbyn
et al. 1993). Packs with more than 20 wolves were not
uncommon in WBNP (Carbyn et al. 1993), although the
annual average pack size ranged from four to 15 wolves
(Carbyn et al. 1993)

Moreover, given the huge amount of bison carrion
made available by other sources of mortality, particularly
hide hunters who took only the skins, leaving behind the
rest, wolves likely scavenged on bison more often than
they hunted them, and this free meat may have led to
large pack sizes. A recent summary of contemporary
studies determined that wolf litter sizes increase an average of 31% with a sixfold increase in ungulate biomass
available per wolf (Fuller et al. 2003). Because bison are
difficult and dangerous to kill, scavenging has likely always been the predominant means by which wolves uti-
lize bison. For example, in northern YNP, more than 50%
of the bison fed on by wolves were scavenged rather than
killed. By contrast, elk are more often hunted than scavenged by wolves (Metz et al. 2012). Historical accounts of
wolves scavenging on bison killed by human and nonhu-
man sources support the hypothesis that scavenging was
a common way that wolves utilized bison. However, the
great wave of hide hunting that started as early as Catlin's
1832 visit to the Upper Missouri, and the large number of
bison carcasses it yielded, calls into question whether the
apparent abundance of wolves and super-large pack size
was natural or an artifact of human activity.

Although the spectacle of a large pack of wolves
hunting bison suggests a high degree of cooperation,
only recently has enough data accumulated to test this
hypothesis. MacNulty et al. (2014) analyzed over 200
wolf-bison encounters from Yellowstone National Park,
including those described in accounts 3–7 of this chapter. They found that pack capture success increased by
40% with each additional pack member up to nine to
13 wolves. Beyond this threshold, pack size had no apparent effect on capture success. These results imply
that wolves in packs as large as nine to 13 cooperate to
capture bison whereas some in larger packs hold back
or "free ride." The reason to free ride is highlighted in
Catlin's portrait of a wolf-bison interaction (fig. 6.2): bison are dangerous. But, in smaller packs, this danger is
outweighed by large improvements in capture success

FIGURE 6.8.   The conditional nature of bison mutual defense. Bison commonly defend one another, but only when the risk of predation to themselves is low. In the example above, snow around where wolves have captured a bison deters others from helping it due to the difficulty of defending against wolves in snow. (Photo by Doug Smith.)

with each wolf's participation due to the small chance a solitary wolf will capture a bison by itself. In general, low solitary capture success promotes cooperative hunting because it leaves ample scope for an additional hunter to improve the outcome enough to offset its costs of participation (Packer and Ruttan 1988; Scheel and Packer 1991). Thus, wolves in smaller packs have a strong incentive to cooperate when hunting bison despite the risks of injury and death.

The notion that the fitness cost of death is far greater than failing to make a kill represents a slight twist on the old maxim called the life-dinner principle: if a predator fails to kill its prey, then it only loses a meal, whereas if the prey fails to escape the predator it loses its life. Thus, it is argued, the force of selection is stronger on the prey than it is on the predator (Dawkins and Krebs 1979). But in the case of wolves hunting bison or that of any other predator hunting dangerous prey, the imbalance between predator and prey with respect to the penalty of

failure is the possible death of the predators. As a result, there can be strong selective pressure for minimizing the risk of prey-caused death during predatory behavior. Free riding during a group hunt is one such behavior (MacNulty et al. 2012).

Another cautious hunting behavior is to attack prey only when they are vulnerable. In YNP, wolves only attacked bison when it was safe to do so. In winter, this involved attacking bison that were caught in deep snow or on a small patch of bare ground with minimal defensive space to maneuver against wolves. Wolves rarely attacked bison on large, open patches of bare ground, particularly if the herd was large.

The mutual defense behavior of bison is a similarly risk-sensitive, context-dependent decision. Specifically, bison only defend each other when it is safe to do so. In YNP during winter, this involves coming to each other's aid when on open ground but fleeing from wolves without regard to any other bison when they are caught in deep snow. Another example is when wolves are latched onto a bison, and it is dragging them through the snow to the safety of bare ground and the bison there. Interestingly, the bison already on the patch do not leave its safety to rescue their comrade (fig. 6.8). Rather, they wait until it has reached the very edge of the patch, where they are themselves standing on bare ground, to strike at the wolves. Both historical and contemporary accounts emphasize the cooperative defense behavior of bison. But the most recent accounts of wolf-bison interaction in YNP demonstrate that mutual defense among bison has it limits. That is, bison engage in mutual defense only when it does not risk their individual safety to do so (Carbyn et al. 1993; MacNulty et al. 2001).

7

Musk Oxen

AT FIRST GLANCE it might seem that the musk ox would represent one of the hardest prey animals for wolves to kill. Large and shaggy and sporting a huge blocky head with sharp, recurved horns, the musk ox appears formidable. Groups of musk oxen form defensive rings or lines, with each animal facing outward and any calves behind. To confront such a facade would seem to a human to be foolhardy. So, too, to many wolves.

Musk oxen originally inhabited, in recent times, northern Alaska, the far northern periphery of Canada west of Hudson Bay, the Canadian Arctic Islands, and the northern and eastern periphery of Greenland (Nowak 1991), but they have also been successfully introduced to several other high-latitude areas of the world (Gray 1987).

Weights of adult musk oxen vary considerably, depending on where the animals live. Average adult male weights vary from 262 kg on Devon Island (Hubert 1974) to 341 kg on mainland Canada (Tener 1965), and average female weights from 172 kg (Lent 1978) to 198 kg (Latour 1987). Captive bulls, however, can reach 650 kg and cows nearly 300 kg (Lent 1978). Newborn calves weigh 10–14 kg. Both sexes sport sharp, recurved horns, which are readily used to hook attacking wolves. Mech has seen a wolf mandible that had the tip of a musk ox horn embedded in its inside aspect.

Even the long, shaggy hair of the musk ox affords the animal some defense against wolves. Some guard hairs measure 62 cm long (Lent 1978). A wolf trying to bite a musk ox while both are whirling around would have to find the actual body through all that fur and must also contend with mouthfuls of shag in the process.

Musk oxen spend most of their time in groups, which confers many of the same antipredator advantages as in other wolf prey (chap. 1). Musk ox herds usually contain adults of both sexes and young, but same-sex groups also occur. The best information about musk ox herd composition comes from Gray (1987, 91): "Herds usually contain both males and females, but single-sex herds composed of all males and, occasionally, all females are also seen. Out of 94 single-sex herds seen between 1968 and 1978, only 5 contained females only. Herd size ranged from 2 to 8 for males, and 3 to 10 for females. Single-sex herds were seen between May and October, with the highest number in July, August, and September. Most herds were composed of adults, though subadult bulls were present in a few of the all-male herds."

Musk ox herd sizes vary by season, with the largest occurring in winter (Hone 1934; Gray 1987; Heard 1992). On Bathurst Island, herds usually varied in size from three to six musk oxen during summer and five to 15 in winter but sometimes reached as many as 60 in winter (Gray 1987). The largest herd reported was 100 on Queen Maud Gulf and on Melville Island, and there is a strong correlation between herd size and wolf density (Heard 1992). Solitary musk oxen are usually bulls, which can drift in and out of various herds (Gray 1987).

One of the benefits of herding is that it provides the group with the sensory abilities of each member. This advantage may be more important for musk oxen than for other prey because musk oxen senses do not seem to be very sharp. Although Mech has not done any formal testing of these senses and little other information is available, he has observed and interacted with musk oxen many times on Ellesmere Island and a few times near Aylmer Lake in the Northwest Territories. His general impression is that musk ox eyesight is poor. The animals generally do not begin grouping up until wolves get to within 100–150 m (accounts 3, 13, 17) but sometimes when wolves are as far away as 400 m (account 25).

FIGURE 7.1. When confronted by wolves, musk oxen face them. (Photo by L. D. Mech.)

It appears that musk oxen might not actually perceive approaching animals as wolves but only respond to any moving creature the same way. Mech once saw a herd group up on spotting four white arctic hares running some 500 m away. At the other extreme, Gray (1987) watched a single wolf surprise a single prone bull by poking his nose into the bull's rump (account 4).

Like caribou, musk oxen tend to drift around over large areas. Although this behavior might be necessary to forage on sparse vegetation that characterizes much of the creature's farthest northern range, it also helps minimize the likelihood of encountering wolves, a considerable benefit to the musk oxen. Little is known about the specifics of musk ox movements except in areas where they have been introduced. The animals are capable of long travels; some have wandered as far as 300 km after having been transplanted to a new area (Lent 1978). In their native environment, musk oxen may move in highly variable ways. On Ellesmere Island, Mech noticed that herds during summer often disappeared after be-

ing in an area for a few days. His general impression was that individual herds drifted around areas as far apart as 50 km. In the Aylmer Lake area, musk oxen seemed to drift over areas as far apart as 10 km over periods of a few days.

Aside from their nomadism making them harder for wolves to find, the primary defensive strategy of musk oxen is to face any attacking wolves (fig. 7.1). When not threatened by wolves, a herd will graze for hours on end with members being spread out as far as 50 m from one another and members of a herd covering an expanse of 200–300 m. At the slightest kind of disturbance, however, herd members stop grazing and become alert. If they confirm a disturbance, they instantly assemble into a tightened group (fig. 7.2). They pack themselves together so tightly that their backs and rumps press against each other, constantly assuring them that no space occurs among them (Gray 1983). Any calves huddle right behind the adults, and if the herd is large enough it forms a circle with the calves inside. Otherwise the adults form

FIGURE 7.2. As wolves persist, musk oxen tighten their group.(Photo by L. D. Mech.)

a line or semicircle with calves behind, adults trying always to face out against the wolves, similar in fashion to bison defensive behavior.

During interactions with wolves, musk oxen defenses are fairly stereotyped. Gray (1987, 129) studied them in detail, based on 20 attacks by wolves that he observed (fig. 7.3). Gray found that

sometimes encounters do not progress beyond the wolves' initial approach to the herd, and the musk oxen do not group together in the defence formation. More often, once the approaching wolves are perceived, an alarm snort is given and the herd groups together. The herd may stampede before forming a defence line or circle, and sometimes stampede after taking up a defence formation.

Once in a defence formation, most musk oxen, including cows, bulls and subadults, gland-rub when wolves approach to within fifty metres. The lead bull gland-rubs more vigorously, sometimes horning the ground, or rubbing his preorbital glands against the ground [thought to indicate a threat (Gray 1987)].

In most interactions several different musk oxen (bulls, cows, or subadults) charge the circling wolf or wolves. Charges by single musk oxen facing one or more wolves may occur, with over twenty three charges during a single interaction. Occasionally, part, or all, of the herd charges at the wolves together. Sometimes a group follows close behind as another musk ox charges, preventing the wolves from cutting off the charging individual.

This aggressiveness on the part of musk oxen, combined with the damage these formidable animals can do with their hooves and horns, most often deters wolves from pressing their attacks. Most of the time, then, the combination of nomadism, grouping, and aggressiveness work well, and unless a musk ox is a calf, an old individual, or debilitated in some way, it survives attacks by wolves (Tener 1954a).

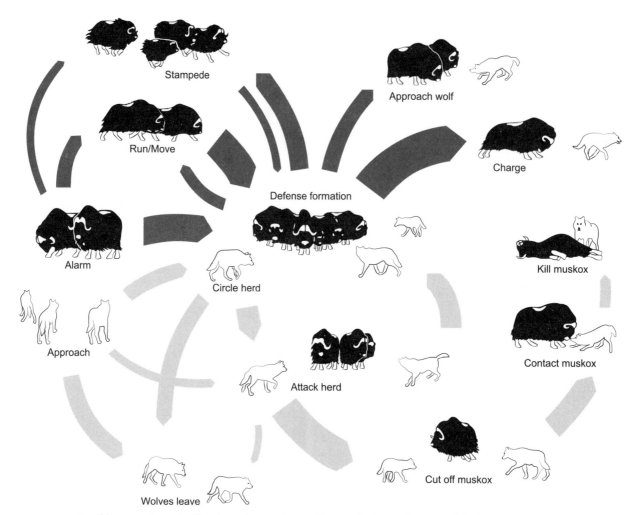

FIGURE 7.3. Possible progressions of wolf pack attack on musk oxen (Gray 1987). Diagram (*starting at left*) showing the more common sequences of events in 21 observed encounters. Order of events shown in the lower half for wolves (*paler arrows*) and in the upper half for musk oxen (*darker arrows*). The wider the arrow, the more often the sequence occurred. (Redrawn from Gray by Kevin Schuhl.)

Hunting Accounts

1. June 20, 1951—Eureka Area, Ellesmere Island, Canada (Tener 1954a).

A herd of 14 musk-oxen that had been feeding undisturbed for several hours formed a defensive group. Two wolves, one white and one gray, were lying down together 50 m from the herd. Occasionally one of the wolves circled the herd and then returned to lie down. Eventually 10 of the musk oxen lay down, while four remained standing facing the wolves. The calf in the herd kept close to the cows, grazing near the resting adults until the white wolf suddenly dashed around the four standing adults and toward the calf that was now outside the group of lying animals. The calf immediately ran to

the center of the herd, and all the musk oxen arose. The adult bull charged the wolf in an attempt to gore it, but the wolf nimbly turned aside and trotted off to its mate. Both wolves left the vicinity about half an hour later, heading toward the east end of the fjord.

2. May 28, 1968—8 km West of Goodsir Inlet, Bathurst Island, Northwest Territories, Canada (Gray 1970)

A lone, bull musk ox could be seen feeding on a riverbank in the broad valley about 1.6 km from a herd of 11 musk oxen. At 2020 hr, I saw a wolf galloping toward the bull from about 100 m away. The wolf ran swiftly along the snow, following the slight depression of the snow-filled riverbed in its approach. The wolf ran close up behind and to the right of the bull, stopped suddenly, as yet unnoticed, hesitated, then darted aside as the musk ox wheeled around and charged. There followed a series of charges with the wolf nimbly advancing and retreating, circling around as the musk ox attempted to make contact. After several such charges the musk ox seemed to collapse into a sitting position with lowered hindquarters. This posture lasted only a few seconds, and the bull rose again; meanwhile the wolf remained standing motionless in front of the bull. The bull continued charging, wheeling around and running at the wolf, forcing it to run out ahead to avoid the horns.

By 2030 hr, there were brief periods when both animals stopped and stood facing each other for several seconds. At times, the musk ox seemed to be slowing down and would back away from the wolf rather than charge it. Suddenly, the wolf grabbed hold of the bull's face for a few seconds but was shaken off. The size of the arena became smaller as the wolf circled constantly, the musk ox wheeling around to follow it. Again both stood facing each other, resting this time for 50 sec. At 2047 hr, the wolf again closed its jaws on the face of the musk ox, this time hanging on for 15 sec. Then followed more nimble circling and dodging, with the wolf seizing hold again but only for an instant, and when the animals separated, a great patch of blood appeared on the head of the wolf. At 2100 hr, again the wolf moved in close, keeping low and avoiding the horns by coming straight in, this time grabbing a secure hold on the right eye orbit of the bull. The musk ox charged forward and backed off, swinging his great head vigorously side to side, only after some time dislodging the powerful grip of the wolf. At this separation, the blood could be seen spreading over the right boss and entire face of the musk ox, and over the face of the wolf as well.

The wolf faced the musk ox constantly, bounding in with lowered forelegs, keeping the head low and close to the face and nose of the bull. Since the wolf kept right in front of the bull, each charge ended with a sharp upward toss of the bull's head, rather than the more dangerous sideways hooking of the horns. Each time the musk ox charged forward, the wolf backed up, then instantly followed in again, moving so closely with the bull that they gave the impression of being tied nose to nose.

Several times the wolf stopped and stood looking back over its shoulder or briefly sniffing at the snow before renewing its constant circling attack.

At 2110 hr, as the wolf moved around past the musk ox to the left, the canid suddenly swung around and moved in to the bull's left side, behind and below the horn, and pulled the bull down. The wolf circled around and moved in again with its head at the left side of the throat and its right forepaw up on the bull's shoulder. The musk ox tried to lift himself up, got up onto his fore knees then collapsed. A second time the bull rose onto his forelimbs and again fell while the wolf stood motionless beside him. On the third attempt, the musk ox rose onto his fore knees and wheeled his forequarters around to face the wolf, tossing his head and horns at the blood-spattered wolf. The wolf moved again, circling around, and the bull, still supported by the front legs, swung his head at the passing wolf, then fell, at this point still holding his body upright.

At 2113 hr, the bull's head remained partly lifted although he now lay over on his side; then as the wolf circled around again, the bloody head lifted toward the wolf, then flopped over onto the snow.

3. September 1970 — Polar Bear Pass, Bathurst Island, Canada (Gray 1987)

We watched six wolves, including two pups, approach a herd of 12 musk oxen. Traveling in single file, the wolves approached to within about 100 m of the herd, which grouped, then separated. As it shifted around, one wolf lay down as two others circled the milling herd. As the herd started to run, the dominant bull stopped, turned, and faced the chasing wolves; a wolf jumped up at his flank, and as he whirled around, the rest of the herd moved up behind him. The bulls at the end of the phalanx-like formation turned to follow the circling wolves. The other wolves then joined the first two, and the musk oxen grouped together in a rough circle facing them. The two pups stood together watching as several adults ran around the herd. Two musk oxen charged out at one wolf, which just ran out far enough to avoid them and then moved right back to the herd again. At this point, the two pups headed away from the closely grouped herd, their tails between their legs. A bull and cow each charged as a wolf ran around the herd. Each time a musk ox charged one wolf, another wolf rushed between the individual and the herd but failed to cut off the musk ox because the rest of the herd closed the gap. Four minutes after the start of the attack, the wolves left the herd and headed toward the pups. The herd remained grouped tightly together as all the wolves lay down.

Two hours later, the wolves approached to within about 20 m of the herd. Five wolves (including at least one of the pups) surrounded the herd, spaced out almost evenly. The herd shifted and split sufficiently for one wolf to move to within a meter of a cow's flank before she charged out at yet another wolf. Several times individual wolves nearly made contact as cows ran up to 20 m from the herd toward the circling wolves. After a minute of intense action, the two pups retreated from the herd, and soon the adults were all heading away from the herd in single file with the pups lagging behind.

4. August 1973—Polar Bear Pass, Bathurst Island, Canada (Gray 1987)

A group of us watched from camp as a single wolf approached a herd of 27 musk oxen, moving slowly upwind through a flurry of light snow. Most of the musk oxen were lying, facing into the wind and away from the wolf. We were amazed to see the wolf walk right up to a solitary bull near the herd and poke its nose into the bull's rump. The startled bull jumped up and whirled around to face the wolf. He gland-rubbed as the wolf circled, then charged. The wolf ran briefly, then walked away, yawning. Several bulls at the edge of the herd charged the wolf as it moved toward the herd. For 10 min, the wolf walked around the bulls and even lay down briefly before the rest of the herd reacted. Then several cows with their calves, grouped together watching the wolf while Cornice, the tagged lead bull, continued his courtship activities. The wolf, in a great burst of speed, galloped toward two calves and a cow standing apart from the herd but failed to cut them off from the others. The herd then grouped together and moved toward the wolf. For almost 2 hr the wolf pranced around in front of the herd, running back and forth and in circles. At times the wolf approached playfully, pouncing at the ground, then running in all directions. Several times Cornice charged the wolf, and others in the herd moved up with or behind him. Other bulls and cows charged the wolf, individually or together, as it moved in front of the herd. Other approaches to the wolf included a charge by a group with Cornice and two

yearlings. At each charge, the wolf got up and ran off, then returned to the herd. A group of cows and calves ran from the wolf and clustered behind Cornice. The whole herd grouped and faced the wolf in a long line, several cows and calves at times standing out in front. Twice Cornice spent several minutes courting cows when the wolf lay down. The whole herd stampeded briefly, then some began feeding while others moved away. After a final bout of whirling around in what seemed like play the wolf walked away from the herd and down the valley.

5. September 22, 1977—Polar Bear Pass, Bathurst Island, Canada (Gray 1987)

In the late afternoon, another solitary bull musk ox moved along the riverbank toward the wolves. The bull climbed up the bank near two adult wolves, hesitated, then walked directly toward one wolf that was lying down. The wolf jumped up and moved quickly down the riverbank and disappeared. A pup approached and passed within 50 m of the bull. Though the wolves were still close, the musk ox remained feeding in the area for 2 hr.

6. September 23, 1977—Polar Bear Pass, Bathurst Island, Canada (Gray 1987)

A new herd of 15 musk oxen entered the pass, joining the two herds of 13 and four already there. During a snowstorm, five adult wolves approached the herd of 13 musk oxen, the pups staying on the riverbank apparently to watch. The herd grouped facing the wolves. Four wolves lay down facing the herd, and within minutes, one musk ox was lying down, and the others had resumed feeding. The wolves arose almost immediately, heading back to a new kill and the pups, and for the next 5 min, three of the adults indulged in vigorous play.

Within minutes of their return to the new kill, the adult wolves were off again, for a herd of four bulls was climbing the face of a 200 m high hill in full view. The five adult wolves began climbing the slope, evenly spaced in single file. When the wolves were halfway up the slope, two musk oxen had separated from the others. The wolves acted promptly: two headed directly for them, two headed up to the top of the ridge, and one angled uphill away from the herd. As one of the leading pair of musk oxen turned to move back to the other two, three wolves began running toward them. The four musk oxen grouped tightly facing the wolves now approaching them from above and below. Three wolves lay in front of the musk oxen, and one stayed on the ridge above the herd. After about 4 min, one wolf ran toward the herd, and the musk oxen turned and ran north along the slope. They stopped and in a tight group, faced the wolves, which then surrounded them. Then four wolves headed downhill away from the herd, the fifth remaining to lie uphill of the herd for another 10 min. As the wolves moved away, the four musk oxen began to separate. Three of the wolves then started to run and chase each other, galloping in circles on the snow banks and romping playfully for 6 min.

7. September 25, 1977—Polar Bear Pass, Bathurst Island, Canada (Gray 1987)

As the adult wolves rested 2 km to the north for about an hour, the herd of 15 musk oxen moved across the hills toward them. The wolves arose and moved across the valley toward the herd. After 15 min, as the musk oxen walked in single file with two bulls several meters behind, one wolf ran toward the herd. A bull began running, followed by part of the herd. The wolves stopped, and the two bulls turned and faced them. The other musk oxen then started to run down toward the river, stopping as the wolves ran to the two bulls. The wolves darted toward the bulls and circled them; the bulls whirled around in an attempt to face the wolves. Both wolves then left the bulls and lay briefly before moving toward the herd again. Two other adult wolves approached, and suddenly all four wolves were chasing the herd over the steep, snow-covered slope of the riverbank. The wolves caught up with an adult female and a yearling musk ox running behind the herd; the yearling kept running, but the cow faced the wolves, and one wolf ran right around her. She turned again and ran down slope toward the herd. As they disappeared, one wolf leaped at the cow's flank. Within a minute, all the wolves were again in sight walking uphill toward the two bulls. One bull gland-rubbed briefly, then charged the wolves. As he charged, other wolves circled behind him and attacked. Between charges the bulls tried to keep together with rumps in contact as the wolves repeatedly darted in and out. After 2 min, there was a break in the attack; one wolf wandered away, and the others stood off to one side. Three wolves moved in once more, leaping at the bulls and jumping away, before breaking off the attack.

8. September 27, 1977—Polar Bear Pass, Bathurst Island, Canada (Gray 1987)

Four adult wolves and one of five pups moved off to the west at a steady trot, leaving the other four pups behind at the rendezvous site. The wolves passed within a few kilometers of two herds of musk oxen with no indication of having seen them. After an hour of steady traveling (11 km), they disappeared. When they reappeared there was a fourth adult, and they were headed toward 17 musk oxen. The pack split up, with one wolf on the ridge behind the herd and two moving across in front. Four wolves moved together down the ridge behind the musk oxen, the pup trailing behind three adults. Five minutes later they all disappeared.

9. July 9, 1986—Eureka Area, Ellesmere Island, Canada (Mech 1988)

About 2245 hr, five wolves were chasing some hares. Suddenly, the wolves sensed three musk oxen 800 m away, grazing on a wet hummocky area. The pack was definitely interested, for they headed right toward the herd. The musk oxen detected the wolves when about 100 m away but remained nonchalant. The wolves lay down, while the oxen grazed 10–15 m apart. Perhaps each group had confronted the other before, for neither seemed that interested in the other.

Some wolves seemed to go to sleep, and it looked like they were just going to keep the musk oxen on edge until someone decided to move. The strategy worked to the musk oxen's advantage. One wolf soon spied a hare on the horizon and headed after it. Two others followed. Soon the three wolves were seriously attempting to capture the hare. When their hare hunt was unsuccessful, one of the other wolves drifted off, while the other remained sleeping around the musk oxen. About 20 min later, she, too, gave up, and the musk oxen eventually resumed grazing.

10. July 15, 1986—Eureka Area, Ellesmere Island, Canada (Mech 1988)

About 1930 hr, seven wolves appeared, threading their way diagonally down a hillside; they had spotted musk oxen. The wolves moved very deliberately, with little lingering, but at their usual 8 km/hr pace. They seemed interested but not too excited. They approached the herd quite casually, and the musk oxen tightened and faced them. Each adult ox continually shifted its rear around slightly, pressing against its neighbor. The wolves stood and watched 3 m away, with one wolf hanging back a few meters. After several minutes, most wolves lay down, although now and then individuals arose and walked around. They did not seem excited.

The musk oxen, however, took the situation seriously. They stood facing their adversaries, heads lowered and calves huddled at their rumps. Nevertheless, they appeared disorganized. Single oxen on the edges eventually broke rank, and some seemed more interested in eating. Neither were the wolves organized. At one point, the yearling wolf and one of the adult males started strolling away from the musk oxen, while another adult male on the other side of the herd challenged the beasts.

As the casual confrontation continued, however, wolves prowling around behind the herd seemed to unnerve the musk oxen. Gradually, the situation changed into one in which the oxen were more scattered, and the wolves walked about between subgroups. Every now and then a skirmish developed when an ox charged a wolf, even though other wolves and musk oxen just stood around nearby.

Such maneuvering and skirmishing interested the wolves more, and soon they became more active. The unevenness of the terrain may also have played a role, for when a musk ox ran, it had to maintain its footing while carrying its great weight on spindly legs in and out of the tundra troughs without stumbling. Whatever the case, the pace increased, and wolves actually began chasing subgroups of musk oxen. As soon as they got close, the musk oxen would swing around, stand, and threaten. However, with seven wolves chasing various groups, and the groups trying to keep together, the entire herd began gyrating.

The wolves grew increasingly excited. Soon they darted in and out among the frantic musk oxen, often passing within centimeters of them. The shaggy musk oxen would turn and charge with lowered heads, and strike out with front hooves. Fourteen musk oxen and seven wolves formed a swirling, chaotic, dusty mass.

The herd was reluctant to run far, so the attack turned into a localized harassment, back and forth on the flats, up and down through the troughs. The wolves clearly handled the terrain better. Once a musk ox even fell over and lay on its back with all four feet in the air, and the

wolves rushed in and nipped at it. It is hard to say how long the skirmishing went on, but it probably lasted an hour or more, and the pace kept increasing. The herd panicked.

Thirty seconds later, the breeding male and female closed in on a calf, and the female grabbed it by the right side of its head. The male latched onto its nose. The rest of the pack quickly gravitated to the pair and their quarry, while the calf's mother joined the stampeding herd. As the calf struggled, it gradually dragged the six wolves stuck to its head and shoulders down a slope.

Then suddenly, an adult male wolf, which had the most posterior grasp on the calf's right side, let up and rushed off after the herd, which, now in a complete rout, was fleeing down off the plateau into a creek bottom. An adult female, which had the last hold on the calf's other side, soon left to join him. They hit a second calf crossing the creek.

The herd turned right and rushed down the creek bed 100 m to a river. One big bull stood its ground. More wolves left the first calf, while the breeding pair held their grip. The calf was no longer on its feet, and soon the male left. Down in the creek, an adult male and female worked on the second calf, as their pack mates joined them. And in the distance, the remaining 12 musk oxen, including a third calf, were fleeing up the side of a huge bank toward another plateau. The calf and a couple of adults were far to the rear. Then 50 m behind the musk oxen, far away on the hillside, ran another wolf. It was after the third calf! The calf was running along the right side of an adult, and another adult was 5 m behind. The wolf caught up quickly, and just at the top of the hill, grabbed the calf from the right. The musk oxen continued to run, except for the two with the calf. But the closer musk ox did not face the wolf, and the other had stopped partway up the hill. The breeding male wolf arrived and joined the female. The calf continued to struggle and eventually dragged both wolves all the way back down the hill to the river. There they were joined by others, and the third calf fell.

11. July 31, 1987—Eureka Area, Ellesmere Island, Canada (Mech 1988)

A large, lone musk ox, presumably a bull, was grazing on a hillside when six wolves approached. Within 1–2 min, the wolves spotted the musk ox and headed for it. However, like so many moose I have seen, the creature stood its ground and defied the wolves. They circled around it cautiously while it continued to try to face them with lowered head. The wolves quickly lost interest.

12. July 31, 1987—Eureka Area, Ellesmere Island, Canada (Mech 1988)

Five wolves were traveling up a gentle slope toward two large and one medium-sized musk oxen, which had seen them and were standing their ground. The wolves went right up to the three and lay down. The middle-sized ox wedged itself tightly between its companions, and they just stood there. So, too, the wolves. Eventually a few wolves arose, and the musk oxen shifted around nervously. With only three, they really couldn't protect their rear ends, and that's what the wolves seemed interested in. The wolves circled behind, and the musk oxen whirled. More circling and more whirling occurred. Suddenly, the musk oxen panicked and fled, and the wolves shot after them. Whenever the wolves would nip at their heels, though, the oxen would stop and group up again.

All of the musk oxen seemed to know just when one had been contacted too closely by a wolf. I wondered whether there was some sort of sound they uttered when a wolf nipped them, for even the forwardmost musk ox would instantly stop when a wolf got too close to the rear one. The attack, which lasted from about 1245 hr to 1300 hr, continued like this, with the musk oxen trying to remain grouped up, usually on a little rocky knoll that protruded out of the slope, and the wolves trying to get them running. Once when the musk oxen stopped, the wolves rushed in, but one of the larger oxen quickly charged them and dampened their enthusiasm. Finally, the wolves seemed to tire of the attack. I really doubt that they drew any blood. The long, dry, shaggy fur of the musk oxen was probably all they tasted.

13. July 31, 1987—Eureka Area, Ellesmere Island, Canada (Mech 1988)

The instant the six wolves spotted a herd of about 15 musk oxen, including calves, in a valley, they froze. Musk oxen may not be able to see very well, for we—six wolves and two humans—were all out in the open only 150 m or so from the herd, yet the oxen still had not noticed us. Each wolf, still upright, not crouched,

crept forward on point slowly and deliberately. Each gradually inched forward for about 2 min. Suddenly, the breeding pair shot forward. The musk oxen quickly gravitated together while the whole wolf pack charged. The wolves swarmed around the swirling herd but were unable to grab a loose calf. Still the herd seemed unnerved. The oxen started heading uphill. Instantly, the wolves darted in, catching the herd on a steep, uneven slope. The wolves charged right into the panicked herd, and the herd kept running, with scattered members spread out across a broad front, scrambling uphill and trying to keep their footing.

The wolves chased, and the herd ran. Then one of the wolves homed in on a calf and grabbed at it in flight. The calf dropped back, and for a moment its mother hesitated. However, the herd continued on, now up a gentler slope, with the wolves constantly darting in and out and charging them. The musk ox cow was torn between trying to save her calf and seeking the safety of her retreating herd. She chose the herd, similar to bison behavior.

The calf was now on its own with breeding male attached. The pair floundered around over a ridge almost out of view for awhile. Meanwhile, uphill from us another drama was unfolding. The wolves had grabbed a second calf and were dragging it away from the herd. Now, however, the herd was on higher, more level ground, and it was grouping up. A cow charged out of the herd and challenged the wolves tearing at her calf.

The wolves let up for a moment, and the calf shot toward the herd. Again the wolves grabbed it, but again the cow beat off the wolves as both oxen neared the safety of the herd. The wolves dashed in once more and made another try for the calf just as it began intermingling with the adults. The cow joined in the group again, and the wolves came flying out.

As soon as the herd had regrouped on a high and level spot, the pack again gave up. Suddenly, an adult female wolf realized that the breeding male had caught the first calf and was still working on it. The wolf shot straight for the struggling pair and joined in, grabbing the calf by the head. The calf bucked, bolted, and bellowed, and that brought the rest of the pack. Each new wolf also went for the head, and they all just hung on. Once the calf bucked hard enough to throw all the wolves and took off. But one wolf grabbed it by a back leg, and the others piled on its head again.

It was unclear just when the calf died, even though it was only about 25 m away. The wolves just kept tugging at its head, and it kept resisting for perhaps 5 min more. Then it fell.

(Footage of a musk ox herd trying to protect its young from wolves can be seen in the video "Musk Ox vs. Wolves," published by *National Geographic*, December 31, 2012, http://www.youtube.com/watch?v=mEYQdCNm9xA.)

14. August 4, 1987—Eureka Area, Ellesmere Island, Canada (Mech 1988)

An adult musk ox happened to wander by the pack's rendezvous site. The breeding female was intently carrying something that she seemed to want to cache when she came up over a rise and confronted the musk ox. She stopped and looked at the ox, and the two just stood about 30 m apart, staring at each other for a few seconds. The wolf then resumed her task, and the musk ox continued on.

Meanwhile, a yearling wolf that was tracking the female also came on the musk ox. The yearling headed straight toward it. The ox, however, quickly charged the yearling when the two were 15 m apart and chased it for another 30 m. The yearling then left.

15. August 5, 1987—Eureka Area, Ellesmere Island, Canada (Mech 1988)

A musk ox herd was standing in defense formation on a little plateau. Two adult male wolves were lying about 100 m in front of them. Following is a chronology of what happened during the next 2 hr.

10:15 hr—One wolf arose and went around west and north of the musk oxen and approached from the north, and the musk oxen tightened their group. There were seven adults and one calf in the middle.

10:18 hr—One musk ox lay down.

10:20 hr—One wolf headed north around musk oxen.

10:25 hr—Two musk oxen lay down.

10:32 hr—The south wolf approached the herd to within about 30 m and then went around to the north.

10:33 hr—All musk oxen lying.

10:34 hr—Both wolves gave up and headed north, then east, while the musk oxen pressed together.

Eventually, their shifting increased, and after a few minutes, they began ambling. Gradually their pace increased. They seemed like they were trying to steal away. The breeding wolf pair arose and halfheartedly started toward them. However, this new charge was only token; the pair quickly aborted it. The wolves then continued to rest for another 45 min and left.

16. June 27, 1989—Eureka Area, Ellesmere Island, Canada (L. D. Mech)

The breeding male seemed very intent on heading west, and four others followed but lagged toward a musk oxen herd. But the herd was alert. The wolves, which had now all caught up and were 30 m east, were sneaking intently toward the musk oxen and peering over the hillside at them. The herd grouped. The wolves watched a few more seconds, saw the grouped-up herd, and then gave up.

17. June 27, 1989—Eureka, Ellesmere Island, Canada (L. D. Mech)

At 0305 hr, a pack of five wolves arose and approached the top of a small plateau where a herd of 12 adult and one calf musk oxen were feeding about 150 m north; they were aware of the wolves. A large male musk ox moved toward the wolves about 30 m while the herd grazed. Twice the herd rushed together in defense formation but then grazed again. The wolves remained on the plateau. At 0347 hr, the entire herd joined the male approaching the wolves. They moved about 75 m toward the wolves, and the wolves moved toward them in ones and twos. The wolves seemed more curious than interested in the musk oxen, moving closer and walking within 20 m. When the last wolf had walked by, the musk oxen chased it, and the wolf ran. The wolves left at 0420 hr.

18. July 4, 1989—Eureka Area, Ellesmere Island, Canada (L. D. Mech)

At 0808 hr, while the breeding male and five offspring wolves slept on a small hill, seven adult and one calf musk oxen walked within 30 m and went right on by. Every now and then a wolf seemed aware of the musk oxen but did not move. I thought it was a ploy, but the wolves just were not interested then. At 1240 hr, the wolves left north, and I suddenly found two had surrounded a grouped-up herd of eight adult and one calf musk oxen but gave up in 1–2 min.

19. July 4, 1989—Eureka Area, Ellesmere Island, Canada (Mech 2011)

At 1320 hr, six wolves switched direction and headed southwest up a hill where I could see several musk oxen herds 1–2 km away. The wolves quickly got to the nearest herd (nine adults and two calves) 800 m away, and at 1325 hr looked intently at the herd from about 200 m, but the herd was already tightly grouped. Both wolves and musk oxen lay down 1325 hr–1405 hr. Then the musk oxen grouped tighter, and the wolves left.

20. July 4, 1989—Eureka Area, Ellesmere Island, Canada (Mech 2011)

Six wolves were on a flat with a group of four cow and three calf musk oxen grouped very tightly, and about 200 m away was a herd of nine cows and three calves. In between, about 50 m from the smaller herd, stood a calf apparently confused about which way to turn. The six wolves sped by the larger herd and headed straight to the calf. They all then grabbed the calf around the head as usual and pulled it down (1422 hr). The wolves killed the calf within about 5 min.

Then I noticed about 50 m straight ahead (the rest of the action was taking place to the right and ahead) what turned out to be a yearling musk ox already wounded in the nose. The wolves came back and worked on it, grabbing it mostly by the hind legs until they had it down. They killed that animal within about 5 min also and began feeding on it (1437 hr).

After killing the yearling and calf and running back and forth between them, the wolves noticed a single bull that I had seen "hiding" in the creek gully, which was about 25 m wide and 4 m deep, run up the bank and toward the herd of nine cows and three calves. The wolves headed straight toward him and tried to attack, but the bull whirled and charged them, and within 30–60 sec the wolves had given up; the bull got into the herd. Two other adult musk oxen elsewhere in the gulley also came up and joined the larger herd. The wolves tried to attack them also but gave up quickly.

The whole herd (now nine cows, three calves, two adults, and one bull) then began gradually to drift away from the wolves, which were feeding about 150 m away. As the herd moved away, individuals became anxious and began to move faster. Soon the whole herd was running toward the gulley. Instantly, the wolves stopped feeding and headed straight toward the herd, which fled across the gulley and then stood on top of the east bank and fought off the wolves.

(My whole impression of the overall area is that the terrain helped the wolves. Perhaps they used the gullies to get close to the musk oxen. There must have been quite a bit of surprise and confusion when the wolves arrived, since the calf had been left out alone, the bull was caught alone in the gully, and the two other adults were in the gully, all separated from the two herds.)

21. July 5, 1989—Eureka, Ellesmere Island, Canada (L. D. Mech)

The pack of six (account 20) left the calf remains at 2022 hr and headed east then northeast out of sight until 2035 hr when they headed northeast toward two musk oxen herds and a single out on some broad flats. At 2045 hr, they charged a herd of five adults and one calf, which grouped up. The wolves left.

At 2050–2100 hr, the wolves charged directly at a herd of 10 adults and two calves from at least 200 m away, but the herd grouped up, and the wolves gave up in a few minutes.

They then headed to a single musk ox and chased it for about 5 min, half the time out of our view, but then gave up.

At 2115 hr, the pack chased two adult musk oxen about 2 min, gave up, and continued northeast.

22. July 6, 1989—Eureka Area, Ellesmere Island, Canada (L. D. Mech)

At about 1600 hr, the six wolves (accounts 20 and 21) spotted a herd of about 10 adults and three calf musk oxen and immediately started chasing them. They caught a calf right in front of us and wounded another. The musk oxen were quite antsy and ran easily, and the wolves chased them. The herd went northeast out of sight, and the whole pack followed after eating all the innards out of the calf. The calf was a male about 30 kg.

23. July 13, 1989—Eureka, Ellesmere Island, Canada (L. D. Mech)

A subordinate female wolf veered off her course to walk by six adult musk oxen. They got up and faced her but didn't group. The wolf continued north.

24. July 6, 1993—Eureka Area,
Ellesmere Island, Canada (L. D. Mech)

A breeding wolf pair passed within about 50 m of two male musk oxen but showed no interest in them.

25. July 12, 1996—Eureka Area,
Ellesmere Island, Canada (L. D. Mech)

A breeding male wolf spotted six adults and two calf musk oxen across a valley about 1 km away, and he and the breeding female headed for them. At 2100 hr, when the wolves were 400 m away, the musk oxen grouped. The female wolf, followed 30 m by the male, headed right on by the musk oxen. At one point, the musk oxen lined up on top of the hill where they had grouped up and looked down at the wolves—2102 hr. The breeding female, followed by the male, continued heading up the creek beyond the musk oxen.

26. July 8, 1998—Eureka Area,
Ellesmere Island, Canada (Mech and Adams 1999)

At 0200 hr, a breeding pair of wolves that had produced no pups this year spotted three musk oxen about 500 m ahead in a valley at 0224 hr. The wolves continued toward the musk oxen and when 100 m away ran straight at them. The musk oxen fled some 30 m and headed in a tight group up a steep slope, with the two largest animals (a bull and the other presumably a bull) about half a body length ahead of the smallest, a cow.

As the musk oxen were running, about a third of the way up the slope at 0226 hr, the male wolf grabbed the last one (a cow) by the rump and hung on, and the female lunged toward the cow's head. The cow wheeled around, and the male lost his grip. Both wolves focused their attacks on the head and neck of the musk ox, biting at her nose and neck, sometimes hanging on and sometimes losing grip. The musk ox kept pushing up with her lowered head and horns but did not use her hooves. After about 30 sec of the focused attack, one wolf gained a solid grip on the cow's nose and the other immediately attacked the side of her neck, repeatedly grabbing a new purchase. The cow appeared to struggle little once the wolves had gained solid grips on her.

The two bulls had stopped about 15 m farther up the hill, and one of them suddenly charged down at the wolves that were attacking the cow, sending one of the wolves tumbling about 10 m down the hill. (We could not

see whether contact was made, for the bull charged on the opposite side of the cow from us.) The bull hooked repeatedly at the remaining wolf, which eventually released its grip on the cow's nose. By now, the third musk ox had joined the other two, and they headed back up the hill with the cow tightly wedged between the two bulls. The wolves quickly dashed back after them. Again, one wolf grabbed the rump of the cow, which wheeled to meet the wolf head on. The female wolf then grabbed the cow by the nose, and the male grabbed her by the side of the neck. The wolves kept their grips on the cow for about 30 sec, and at 0231 hr, the cow fell on a flat area of the hillside about two-thirds of the way toward the top and stopped struggling. The wolves continued to tear at her head and neck, but the musk ox did not move.

27. June 30, 2000—Eureka Area,
Ellesmere Island, Canada (L. D. Mech)

A pair of adult wolves, when around 200 m downwind of three musk oxen, stopped and became alert, and the male headed toward them. The musk oxen stood and looked but did not group. The male got to within 30 m, turned around, and left. The female did not even approach the musk oxen but sat 15 m behind the male. The wolves continued on.

28. July 2, 2006—Eureka Area,
Ellesmere Island, Canada (Mech 2007b)

At about 2240 hr, seven wolves jumped two hares but did not chase them (which was unusual) and continued on to a herd of seven adult and three calf musk oxen upwind of them. The wolves disappeared, and then at 2250 hr, two wolves were about 200 m from the musk oxen heading toward them up a shallow valley, slowly stalking, while at least four wolves watched intently from a ridge of rock piles approximately 400 m from the musk oxen. Suddenly, the musk oxen ran to each other, two or three musk oxen that were lying arose, and all grouped up. Apparently triggered by the musk oxen running, all the wolves from among both the waiters and stalkers rushed to the herd, milled around them for about 1 min, then left.

29. July 6, 2006—Eureka Area,
Ellesmere Island, Canada (Mech 2007b)

A herd of about 13 adult and seven calf musk oxen had come into view, and at 1628 hr, male, subordinate Wolf A, when on a ridge, stared intently toward the musk oxen for 1–2 min. He then headed to another subadult wolf (Wolf B) lying about 20 m away below the ridge chewing on an object and "nosed" that wolf. Wolf B immediately abandoned the object, went to where Wolf A had stared toward the musk oxen, and also stared toward them. It appeared that Wolf A had communicated with Wolf B, motivating Wolf B to look toward the musk oxen. Both lay down for a few hours.

At 2025 hr, two wolves, including Wolf A, and a wolf that was either Wolf B or one of two other possible subadults headed to within 300 m south of the musk oxen. The wolves had moved 200 m on the flats toward the musk oxen and stopped, stared toward the herd a few minutes, then backtracked 50 m, headed east 100 m and then back north 50 m toward the musk oxen but parallel to their original route. They then lay down out of sight, 100 m from a green, wet, sedge meadow 15 m wide and 40 m long, still 300 m from the herd. (One adult musk ox stood 200 m southeast of them, possibly having passed from the herd through where the wolves were before the wolves got there.)

At 2034 hr, the two wolves were lying hidden around 200 m southeast of the musk oxen; the musk oxen grouped up loosely, but by 2042 hr had resumed feeding.

At 2104 hr, three adult musk oxen meandered down the meadow near the wolves and when the first was within 30 m, the wolves left south down a trench 30 m and circled west and lay in a hidden spot (rough, uneven terrain with lots of trenches and small, 1 m high hillocks) 15–30 m from the meadow. The musk oxen did not appear to sense the wolves and continued southeast. The herd, 200 m away, lay down. The musk oxen and wolves remained where they were for 3 hr, with at least one of the wolves intermittently lifting its head and watching the musk oxen.

At 2330 hr, the musk oxen began arising.

At 2337 hr, the musk oxen grouped for 30 sec and then opened again and began drifting southeast along the same general route as the first three adults.

At 2342 hr, the musk oxen got to within 100 m of the wolves, which then circled east and then north up a trench and charged the musk oxen from about 50 m. The musk oxen grouped, and the wolves milled around 30 m away for 1–2 min, left, and lay down 60 m from the herd. After 2 min, the musk oxen continued drifting away. The wolves arose, and musk oxen grouped for 1 min and traveled on. The two wolves lay back down and slept.

Conclusion

Probably more than with most other prey species, the wolf's ability to travel far and wide really helps the wolf in overcoming musk ox defenses. This is especially true in the highest latitudes that wolves and musk oxen inhabit. That area is so bleak and sterile that vegetation is sparse and scattered. Ellesmere Island, for example, at latitude 76°–83° north is covered with ice and snow most of the year, and less than half of it supports any vegetation. Thus, musk oxen herds are also scattered far and wide and with their low density and nomadic tendency, they can be difficult to find at any given time. This is the challenge facing any wolves that depend on musk oxen.

But wolves are up to the task. A couple of examples will illustrate. On July 31, 1987, a pack of six wolves being studied by Mech on Ellesmere left its den of pups at 0930 hr and traveled without resting very often most of the day and located a herd of musk oxen that they attacked at about 1500 hr (account 13). The distance they traveled is unknown, but the straight-line distance between the musk ox they killed and the den was 32 km, and the wolves had traveled a kind of semicircle to get there. After they fed, some of the wolves then returned the 32 km back to the den to deliver food to their pups (Mech 1988).

A second example, also from the Ellesmere wolves, involved a pack of at least 20 in winter (Mech and Cluff 2011). This pack traveled straight-line distances of up to 76 km in 12 hr, and covered an area of 6,640 km² over an 8.5-mo period. The pack once made a 10-d foray to the southeast, covering a minimum of 263 km. On its return trip, it averaged traveling a minimum of 41 km/ day. These kinds of travels help wolves locate musk oxen even though musk oxen are nomadic and sparsely distributed.

As with bison, musk oxen usually inhabit open plains, and possibly reflecting this similarity, the defenses of both bison and musk oxen are much the same. They group up to confront wolves collectively but only when their own safety is not at risk (account 13). Given that musk oxen are more closely related to sheep and goats, which depend on steep terrain for their safety, the parallels between bison and musk ox defensive behavior could result from convergent evolution.

Finding a musk ox or herd is only the first step leading to a wolf's dinner. The next step—catching and killing one of the animals—is harder. Wolves are extremely aggressive and persistent when attacking their prey, and they are quick and agile in avoiding hooves and horns. Gray (1987, 129–30) who analyzed 20 attacks by wolves on musk oxen, characterized wolf approaches to musk oxen as follows:

The approaches of wolves toward muskox herds are most often direct with no attempt at concealment; in most the wolves travel in line, but occasionally they split up and approach in a fan formation. Usually wolves approach across or upwind, avoiding detection by scent, and move slowly, stopping to look at each other and the muskoxen. A swift, direct, running approach is not, however, unusual. . . .

Wolves actually attacked the muskoxen in over half of the twenty-one encounters observed by running and jumping at an individual. An attack may be made directly, before circling the herd, or may be preceded by a circling of the herd by one or more wolves.

Wolves are often successful in cutting off an individual muskox from the herd. Wolves usually make contact with muskoxen by leaping at their shoulders or flanks. I observed three encounters where the contact resulted in a kill. Four times, muskoxen touched by wolves escaped without any noticeable injury.

Encounters between wolves and muskoxen can last from two minutes to almost four hours. As many as five periods of rest occurred in interactions during which wolves or muskoxen lay down, sometimes for an hour. Excluding periods of inactivity during an encounter, the maximum length of an encounter was close to three hours. On one occasion, after an encounter of five minutes, I watched a single wolf lie down near a herd for seven hours and forty-four minutes. While the fog rolled in and out, the herd approached the wolf quite closely before moving off, but the wolf did not even get up, nor did it approach the herd again on arising.

Wolves attack musk oxen at whatever point on the body (rump, flanks, head) is convenient, but there seems to be a tendency to grab the head when possible (accounts 2, 26). Even a single wolf killing an adult effectively grabbed its head (account 2).

Mech's observations (several hunting accounts above) parallel those of Gray (1987), except that most of the kills of musk oxen that Mech watched were those of calves in summer (Mech 1988, 2010, 2011; Mech and Adams 1999). One other notable difference is that Mech observed wolves doing more slow stalking (hunting account 13), sneaking (account 16), and concealing their approach in uneven terrain (account 29), possibly even using more complex strategies (Mech 2007b).

When attacking musk oxen herds with calves, the wolves' main technique was to try to separate a calf from the herd by dashing into the middle of the herd, especially whenever the herd panicked and fled (accounts 10, 13, 22). It sometimes took considerable harassment before the herd stampeded (fig. 7.4), but when it did, the wolves became much more aggressive in trying to grab a calf. Once a wolf got a good hold of a calf, by a hind leg or the head, it generally held on until other wolves joined it, usually also grasping around the head and neck, and pulled the creature down.

The extent to which calves form a high proportion of musk oxen that wolves kill is unknown. Gray (1983, 1987), who was in the field year-round, watched wolves kill five bulls and two cows, whereas Mech (2010), who was in the field only during July on Ellesmere, observed wolves killing seven calves, one yearling, and one adult and found fresh kills of four other adults during that period. Tener (1954a) found remains of 26 musk oxen on Ellesmere, and only one was of a calf. However, calf remains are few and disintegrate faster, so the proportion of their remains is not representative of the proportion of deaths. Of the adults Tener found, most were older than 7 yr, and there were twice as many bulls than cows. Once more, these few figures suggest that the same dynamics are at play with wolves killing musk oxen as other species. That is, wolves generally end up with individuals that are easiest to kill with the least risk to themselves (fig. 7.5).

FIGURE 7.4. After considerable harassment by wolves, musk oxen may flee (account 20). (Photo by L. D. Mech.)

FIGURE 7.5. Calves are easier for wolves to catch and kill if the wolves can separate them from their mother or the rest of the herd. (Photo by L. D. Mech.)

8

Miscellaneous Prey

WOLVES ARE OPPORTUNISTS that will eat almost anything. That includes just about any animal and many types of fruits, such as berries and melons (Peterson and Ciucci 2003). Mech (1970, 168) claimed that "the wolf cannot survive on plants." However, that assertion has never been tested, and given more recent information that wolves eat many types of fruit and that many captive wolves are fed primarily on dry dog food, one could be skeptical of Mech's claim. However, we now know that wolves lack certain enzymes for digesting starch (Axelsson et al. 2013), so it seems more likely that wolves cannot survive on plants for long periods.

All that aside, most wolves customarily eat freshly caught animals, usually hoofed creatures. Other types of food as well as carrion tend to help sustain wolves over periods of food shortages when the animals become more desperate. As indicated in the previous chapters, wolves in most areas ordinarily hunt deer, elk, caribou, moose, and other ungulates. Nevertheless, under various circumstances, wolves do hunt, kill, and eat other types of prey. These circumstances include the following: (1) when primary prey are unavailable, scarce, or less vulnerable; (2) when alternate prey are abundant and/ or easily caught; and (3) when wolves are traveling to find, catch, and kill regular prey and happen on other creatures.

This chapter covers these alternate prey species, including such creatures as pronghorn antelopes, wild horses, wild boars, seals, beavers, hares, ground squirrels, waterfowl, salmon, and mice. Not much information is available except isolated anecdotes for most of these species, except arctic hares. More information is available for wolves hunting arctic hares than for the other alternate prey species because in some areas these large hares form a high percentage of the wolf's diet

(Tener 1954a) and because wolf–arctic hare interactions have been studied more than such interactions between wolves and other small prey (Mech 2007b). Nevertheless, for the sake of completeness, we include whatever observations we could find of wolves hunting these other prey.

Pronghorn Antelope

Pronghorns inhabit primarily the western United States and northern Mexican plains. Their range overlaps with wolf range in Wyoming, Montana, and Idaho, including Yellowstone Park, where wolves occasionally prey on them. Only a few written observations are available, along with two video clips on YouTube ("Druid Wolves Chasing Pronghorn," filmed by Chad Shelton, June 18, 2001, http://www.youtube.com/watch?v=cvIz46qtk30; and "Wolf Chasing Pronghorn in Grand Tetons," filmed by Chris Floyd, September 25, 2010, http://www.youtube.com/watch?v=Nf09Lx0-sBk).

1. June 15, 1997—Yellowstone National Park, Wyoming (D. Stahler)

A single doe pronghorn walked hesitantly away from a clump of sage where a fawn lay hidden. At 1925 hr, subordinate female Wolf 17F trotted toward the pronghorn grazing 300 m away. The wolf came within 100 m before the pronghorn snorted, stamped, and trotted to a nearby hill. Wolf 17F circled around toward the sage clump. The pronghorn rushed the wolf, then stopped within 50 m. The wolf trotted toward the pronghorn, which then trotted away. Suddenly the pronghorn wheeled and charged 17F, quickly gaining speed causing the wolf to dodge the pronghorn. The wolf then lunged toward the doe, which fled. The doe soon stopped, glanced toward the wolf and then toward the nearby sage clump. The doe again charged the fleeing wolf, nearly overcoming it. When 2–3 m from the wolf, the pronghorn jumped in the air kicking both front feet up and out in front simultaneously, nearly striking the wolf. The wolf ran at top speed toward the forest edge with the doe immediately behind. When the wolf reached the treeline, the pronghorn stopped, ran back into the opening away from the wolf. Thirty minutes later, the pronghorn grazed her way back to the sage clump, where a young fawn appeared to nurse for 10 min before bedding out of sight again.

2. June 27, 1998—Yellowstone National Park, Wyoming (R. McIntyre)

At 1224 hr, seven Druid Peak wolves were chasing a pronghorn and her fawn at top speed. At 1226 hr, the wolves excitedly stuck their heads into thick sage. Wolf 42 ran from there, carrying an intact pronghorn fawn. The pronghorn mother approached the kill site at 1238 hr and stared at it. The wolves killed two pronghorn fawns there.

3. May 7, 2011—Yellowstone National ark, Wyoming (L. Lyman)

A group of pronghorn were on the move, and the yearling wolves were after them. The pronghorn were zig-zagging as usual when the group turned to go west; one adult pronghorn just didn't quite make it fast enough. She was in deep snow, and just as she tried to turn, a wolf was right there. Grabbing her by her narrow ankle, the wolf had a strong hold and the wet, deep snow definitely worked in favor of the wolf. Continuing the struggle, the wolf moved up to the back of the neck. Having no luck with that tactic, the wolf went back to the leg and foot. About 2.5 min later, a second wolf showed up and then another and another until all five were there and downed the pronghorn. One pulled out the fetus while others divided up the rest of the inventory.

Wild Horses

Wild or feral horses live only in a few parts of wolf range, primarily in Wyoming; in Galicia and the Cantabrian Mountains, Spain (Llaneza et al. 1996; Lopez-Bao et al. 2013); and in western Canada, including some pack horses left to fend for themselves over winter (Carbyn 1974). As with other potential prey, wolves tend to hunt whatever is available locally, and that includes horses. Only a single account is available of wolves hunting horses, however.

July 26, 1970—Jasper National Park, Alberta, Canada (Carbyn 1974)

At 1345 hr, we observed, from a helicopter, 10 wolves attacking 15 horses, one of which had a bell attached. The horses stayed together, and the wolves surrounded them. Each time a wolf moved closer, the horses turned as a unit and rushed the attackers. The wolves immediately retreated. This sequence was repeated several times over a 10-min period. When we returned to the area at 1555 hr, the wolves had left, and the horses were grazing in the same area. A ground check of the horses revealed no evidence of wounds incurred from the attacks.

Wild Boars

Wolves in Eurasia feed a great deal on wild boars (Peterson and Ciucci 2003), which range throughout Eurasia south of Scandinavia and outside of alpine and desert regions. Wild boars weigh 50–90 kg but can reach 350 kg (Nowak 1991). They are fierce fighters, and males possess sharp tusks that protrude from upper and lower jaws and are used defensively. Wild boars produce litters up to 14, and females and offspring live in groups of up to 20. Science knows little about the nature of wolves hunting wild boars. Probably piglets are most often taken, and sometimes wolves ignore wild boars when, to an observer, they would seem to have a good chance to kill them (Reig 1993). The only information available about actual wolves interacting with wild boars is from a video on YouTube ("Two Wolves Attack Wild Boar," uploaded October 5, 2009, http://www.youtube.com/watch?v=h8Zbjz76iIs). The video appears to be professionally filmed so could have been staged. Nevertheless, it is informative about how wolves attack and try to kill wild boars.

The video purports to show Russian wolves attacking two young male wild boars in summer. There are several cutaways and scene changes, so one cannot be sure of the actual sequence of events. Three to five pups about 6–8 wk old appear with the adult wolves before the attacks. The two boars try to stay together and charge the wolves, which flee on being charged. When the wolves try to grab one of the boars, the wolves have much trouble getting a firm hold. The fur and hide of the boar seems tough and supple, and each time a wolf tries to bite the back, side, or haunch of a boar, it is unable to secure a grip. One wolf does grab the boar above the right ear and holds on while the other unsuccessfully attempts to grab a moving hind leg or other part of the back end. A dead or dying boar is shown toward the end, but there is no scene of the wolves bringing the boar down.

Seals

At times, arctic wolves prey on ringed seals, but the conditions that allow such predation are uncommon. Even rarer are situations in which scientists have been in a position to observe any seal hunting that wolves do. Seals, whose weight averages 49 kg, live primarily in the sea, and in the arctic are mostly locked under the ice for 9 mo of the year (Nowak 1991). They do "haul out" onto floating sea ice in summer, and during winter they use their flippers to keep breathing holes open through 1–2 m of uplifted ice. They also hollow out lairs above the ice but beneath the snow, and females produce their pups there (Smith and Stirling 1975). As the snow melts, seals come up and rest near their holes or along shores. Experienced wolves learn that they sometimes can cash in on these large, fatty creatures or on their pups. Following is the only account of such an attempt.

May 24, 1978—Popham Bay, Lemieux Islands, Northwest Territories, Canada (Smith 1980)

At 1050 hr, a lone wolf moved onto the sea ice. It was probably a male, judging from its stretching posture when urinating. He was out of sight by 1100 hr, paying no particular attention to 11 ringed seals hauled up on the ice, although two seals were scared back down under the ice by his presence.

From 1505 hr to 1655 hr, the wolf stalked and scared away three single and four pairs of ringed seals near their breathing holes. The wolf approached these seals slowly, stopping when they raised their heads to look. We could not tell how close the wolf got to the seals, but twice he appeared to get to within 20–50 m before the seals dropped down into their holes. As the wolf passed

through, two more seals were scared down their holes, even though the wolf did not stalk them.

Beavers

Beavers are large, semiaquatic rodents distributed from northern Mexico through the United States and to northern Canada and Alaska, as well as in scattered parts of northern Eurasia. They inhabit waterways and feed on bark, twigs, and leaves as well as aquatic vegetation. Beaver families build dams of sticks, mud, and detritus to form moats around lodges built similarly. Some beavers also live in holes in banks of waterways. Beavers are vulnerable to predators when on land cutting trees and shrubs for food, and their only defense is to reach deep water. When beavers in the water detect danger, they slap their tails against the water, resulting in a loud report that sends any beavers on land scurrying into the water. Where habitat is poor, beavers must travel farther from water to obtain food and are more subject to predation then. Thus, contrary to the case with ungulates, habitat quality probably determines beaver vulnerability more than does body condition.

Wolves hunt beavers in late winter and spring similar to the way they hunt seals—by catching them away from their holes in the ice (Mech 1966a). We found no actual descriptions of wolves hunting beavers, so we present what is known from reading tracks in the snow. The following accounts are from Isle Royale in Lake Superior, Michigan (Mech 1966a, 152).

During thaws, ice sometimes cracks along docks and islands, leaving crevices which beavers could enlarge. In 1961 when thaws occurred in late February and early March, much beaver activity was evident on March 3. Holes in the ice, trails to trees (some 100 feet [30 m] from the holes) were seen along streams, islands, and docks, and in two of these places, wolves had killed beavers.

The first beaver kill was found near the northeast end of Washington Island. A trail led toward shore from a hole in the ice near a dock, and a few feet out from shore was a large blood-spattered area packed with wolf tracks. . . . A nearby wolf was chewing what appeared to be an everted beaver hide. Two wolves were leaving the island, and four others rested about 2 miles [3.2 km] away. We could not land and verify our aerial observations, but from the sign and the unusual behavior of the wolves, there was no doubt that they had killed a beaver.

On the same day, we found remains of a beaver killed by the pack of three near a small island in Tobin Harbor. The wolves had investigated two beaver houses situated against the island and had found trails leading from them to some fresh cuttings. A few feet from one trail there was a large bloody area covered with wolf tracks. Beaver fur was scattered about, and a well-chewed skull lay nearby.

Wolves sometimes sleep beside beaver runways on the ice (Thurber and Peterson 1993), presumably to ambush the large rodents as they shuffle by to cut fresh food, and R. O. Peterson and D. W. Smith (unpublished) have seen them sleep by beaver holes in the ice and then have found the wolf with the beaver remains there the next morning. Wolves also try to dig out beaver houses, spending much time at it (Mech 1970; Peterson 1977). However, if wolves did break through into a lodge, the beavers would just swim out and away (D. W. Smith, pers. comm.). During the ice-free period, one presumes that wolves also wait along beaver runways or patrol the shores of beaver ponds seeking a fresh beaver trail heading inland. Beavers sometimes travel as far as 90 m from their pond to cut hardwoods (Hall 1971), and the creature's only defense is water, so once trapped on land, the beaver is doomed.

Arctic Hares

The arctic hare is the smallest known consistent prey of the wolf. The only possible exception could be some smaller hare or rabbit or other small herbivore in areas of wolf range that as yet are not well studied. The main reason we even know that arctic hares can be an important wolf prey is because one specific area in Canada's High Arctic, the Eureka Weather Station on Ellesmere Island, is logistically a convenient base from which to study wolves. As early as 1951, J. S. Tener (1954a), investigating musk oxen there, found that 83% of 85 wolf scats contained arctic hare remains. Much later, Mech (2007b) found a strong relationship between summer wolf-pack size and an index to arctic hare numbers in the Eureka area. The following accounts of wolves hunting arctic hares are all from this area.

Tener (1954b) made the first such published observation at 0300 hr in sunlight on August 10, 1951. The wolf was approaching and attacking a herd of 125 arctic hares. The herd of hares grazed over about an acre (0.4 ha) on a hillside when a wolf appeared slightly below and to the left. The wolf singled out an individual at the edge of the herd and chased it through the herd. That hare hopped rapidly on its hind legs much like a kangaroo, and the remaining hares, after the initial alarm, paid little attention to the wolf running through the herd. The pursued hare zigzagged through the herd, then left it to ascend the hill where the wolf caught it just below the top. Tener (1954b) also noted that subsequent observations confirmed that arctic hares hop rapidly on hind legs when closely pursued.

Probably the main reason that in the Canadian far north the arctic hare is a major prey item for wolves is because the wolf's only other possible prey there are musk oxen, and in some areas, caribou. In the Eureka region, however, caribou are few and far between, so arctic hares and musk oxen are about the only consistently available prey.

Arctic hares are distributed throughout the tundra zone of northern Canada, including the Queen Elizabeth Islands, the periphery of Greenland (except the eastern edge), and the island of Newfoundland. The hares weigh up to 6.8 kg according to various authors (Best and Henry 1994), although most probably weigh more like 4.8 kg. They are white for most of the year, and on Ellesmere Island they remain white summer and winter. This oddity no doubt reflects the fact that snow can cover the ground there at any time of the year. When startled, arctic hares stand on tiptoes and look around. They also often run upright, although when pressed drop to all fours.

Arctic hares are said to be able to run at 64 km/hr (Van Gelder 1969), but no documentation was provided for this claim. Two similar species of jackrabbits have been timed at 56 and 72 km/hr (Garland 1983), and R. O. Peterson measured the speed of an arctic hare at 60 km/hr (Mech and Peterson 2003). Thus, 64 km/hr is probably reasonably accurate. That speed happens to be just about the same speed as wolves can attain (Mech 1970). However, besides being able to run quickly, hares can zigzag sharply, much more sharply than can a wolf, which gives hares a distinct advantage while being chased by their much longer-legged pursuers (fig. 8.1).

FIGURE 8.1. Arctic hares can run about as fast as wolves but can zigzag more sharply. (Photo by L. D. Mech.)

Arctic hares produce litters averaging 5.5 young (one to eight offspring) in late spring, and at least some might produce two litters (Best and Henry 1994). During May, June, and July, when leverets are young (probably just June and July on Ellesmere), most hares are spread out over the tundra. Leverets are grayish brown and match their surroundings well. As they grow over the summer, they gradually become lighter until by late summer, when about three-fourths grown, they have become grayish white. Being altricial, the leverets leave the birth nest (a fur and dried-vegetation-lined depression, hollow, or crevice) within a few days (Aniskowicz et al. 1990). While still in their nursing stage, the leverets spread out and hide near rocks, in shallow depressions, or in crevices. They assemble daily at a nursing site, and their mother leaves immediately after they nurse, remaining farther and farther from the site as the young grow and develop. This "spacing out" (Ivlev 1961) is similar to that reported for various ungulates in summer and serves to help maximize wolf (and other

predator) search time and improve survival (Mech and Peterson 2003).

When about 3 wk old, some leverets form nursing bands of up to 20 animals (Parker 1977). This congregating also functions to help thwart wolves and other predators by maximizing predator search time, spreading risk, and confusing any attackers (Hamilton 1971; Nelson and Mech 1981). During late summer and throughout winter, young hares assemble with adults into large herds, sometimes numbering in the hundreds. I have seen some hare herds in mid-July and suspected they might have been herds of adult males, for the females should still have been spread out and nursing young at that time (Parker 1977).

In summary, arctic hare defenses against wolf predation consist of the following: (1) cryptic coloration of young in summer and young and adults in winter; (2) freezing in place by young and adults; (3) high speed and sharp zigzagging by adults and by young older than about 4 wk; (4) spacing out by young and adult females in summer, and (5) congregating in groups by young in summer and by young and adults for much of the year and possibly by adult males in summer; and (6) standing on hind legs by adults, enabling them to see approaching danger earlier.

Hungry wolves, then, must learn to cope with these defenses, and as the following observations made in the Eureka area of Ellesmere Island will show, often the defenses work but not always. Mech and his associates made those observations while studying a pack of wolves that were free ranging but habituated to him and his assistants over a period from summer 1986 through 2009 (Mech 1987, 1988, 1995). Thus he could watch these wolves close up, usually from a four-wheeled all-terrain vehicle and accompany them on their travels (Mech 1994b). The wolves were so tame because they lived so far from human hunters, including Inuit (formerly Eskimos), that they never experienced negative interactions with humans.

The following accounts are lightly edited versions of field notes. Because Mech could identify the wolves individually, he gave many of them descriptive names. Unless indicated as a breeder or an adult (a wolf that had bred during a previous year but not the current year), all named wolves are yearlings or 2-yr-olds. All observations were made in summer, day or night (constant daylight), and all distances are estimates. The behavior of wolves hunting hares during the rest of the year, when the hares are full grown, remains unknown. Also almost all of the observations involve leverets (young hares) of varying ages, with only a few descriptions of wolves hunting adults. No doubt this is because in summer the number of leverets far outweighs the number of adults and because leverets are much easier to catch.

1. July 9, 1986

Five wolves had been lying 100 m from three musk oxen and just continued that way until 1108 hr when one wolf ran southeast toward a hare. Three other wolves followed it. The first wolf ran past the sitting hare, but I could not know whether the wolf saw the hare or not. The wolf stopped about 100 m beyond the hare. The other two wolves ran slowly toward the first wolf, and when they were about 100 m from the hare, the hare stood on hind legs and ran northeast. The wolves on both sides ran for a few seconds, gave up, and lay down.

2. July 20, 1986

Suddenly, a 1–3-yr-old female member wolf put on a concerted run, and I saw her chasing a quarter-grown hare. She chased it 30 sec, caught the hare, and carried it off.

3. July 20, 1986

We then saw the breeding male chase a hare for at least 2 min, when he gave up, looking tired.

4. July 23, 1986

At 1335 hr, a hare appeared on a hilltop 5 m from the closest wolf, which was sleeping near the den. When the hare started running uphill, the wolf awoke and chased it a few meters, and the other wolves awoke and stood up, but the hare disappeared over the hill, and the wolves went right back to sleep.

5. July 24, 1986 (Mech 1988)

At 0409 hr, the breeding female awoke and looked northwest past the den. She had been sleeping 50 m east of the den. She suddenly looked northwest, headed at an excited trot northwest across a stream and up a slope toward an adult hare 400 m away on the top of a hill. The hare did not run until the wolf was 50 m away. The wolf gave a very short chase, then sniffed where the hare had been. Suddenly, two or three small leverets (half the size of a snowshoe hare) sprang off in various directions. The wolf chased one downhill, and I found her eating the leveret at the bottom of the valley, with several other wolves around her at 0425 hr.

6. August 7, 1987

At 2310 hr, Shaggy (1–3-yr-old female) spotted two white but still juvenile hares 200 m west and headed for them. Left Shoulder (2–3-yr-old male) and the breeding male followed. I lost track of Shaggy, but the two males got to within 30 m of the hares and sat there from 2321 hr to 2325:30 hr. They were above the hares on the ridge in plain sight of the hares. One wolf lay down, and both watched the hares intently. Then one wolf went back toward Shaggy, and both headed up toward the hares; the hares ran on hind legs. The wolves probably didn't see them as they did not suddenly rush forward. The hares ran about 125 m, much on hind legs, and stopped on hind legs until 2335 hr. The wolves had given up at 2326 hr.

7. July 17, 1988

At 0400 hr, the breeding female was asleep, but Grayback (1–3-yr-old male) stared intently east toward a hare. Then he started stalking down the ridge, step by step. He put his head and back down and watched. He slowly crept forward, trotted a few steps, and broke into a full run (0404 hr). The hare ran down the ridge, darting left and right, with the wolf behind, turned, and went up the ridge past the female, then turned left, and both disappeared. But several minutes later, I saw them on a far ridge. The hare stayed to rocks on that ridge, going around one pillar on one side, and the next on the other side. Grayback gave up there. At all times, he was 5–6 m behind the hare but could never quite close the gap.

8. July 22, 1988

The breeding male suddenly took off at a full run east down a creek. R. R. Ream then saw a hare running along Devil's Backbone Ridge. But the wolf lost the scent and sight of the hare, spent some time sniffing around, and gave up.

9. July 24, 1988

0911 hr—The breeding female headed southeast away from the area and veered east just south of camp; she jumped a leveret (half grown) and gave a short, almost token, chase, and the hare was about 25 m away and gained quickly; the wolf gave up at about 0925 hr. Around 0927 hr, she jumped another hare only 12 m away, chased it about 10 m, and gave up.

10. July 15, 1989 (R. O. Peterson)

Peterson saw the breeding female again when she was seconds away from catching a leveret. A couple of fast turns was all he saw; then she headed east with the hare in her mouth. She caught the hare in full view of three adult musk oxen, about 50 m from her.

11. August 9, 1989

We watched a yearling wolf chase a half-grown hare for 3 min up and down the hills. The wolf got to within 10 m at times but gradually lost ground and gave up.

12. July 8, 1992

An arctic hare was running about 150 m north of two adult wolves, and the wolves started stalking it. However, it detected them and ran south along the east side of a ridge out of their sight and escaped. Much of its run was on hind legs.

13. July 17, 1992

One of two wolves jumped a half-grown leveret. The hare ran straight downhill toward a fjord 50 m away, and both wolves chased it. There was some zigzagging, but most of the chase was straight, and the wolves caught the hare at the shore. The hare let out a sudden squeal, and each wolf looked up with half the hare in its mouth.

14. July 22, 1992

At about 1105 hr, the breeding male wolf arose as a leveret was running 50 m away. The wolf looked intently and started stalking in that direction, although the hare had disappeared to the north. The wolf continued stalking east and suddenly a hare ran up, and the wolf chased it. Possibly this was a second one since hares often are together, and maybe the wolf knew this. The wolf chased the hare, usually staying 5–6 m behind it but gradually gaining. The hare zigzagged much, but the wolf seemed to be gauging his zigzagging—that is, not zigzagging precisely with the hare. After a minute, the wolf caught the hare, bit it once, let it lie there at 1109 hr, and lay down about four m from it for 5 min before carrying it toward the den.

15. July 22, 1992

At 1835 hr, when the breeding female was within 50 m of where I had seen three leverets together in one spot about 20 hr ago, she became alert and started stalking. Suddenly, three hares headed in three directions. The wolf chased one up a ravine about 300 m but did not get it.

16. July 28, 1992

As the breeding female got just north of two crouching leverets (wind from southeast) around 5 m away, she spotted them, moved toward them intently, and when 2 m northeast of them, turned toward them and approached. They shot out, one north and one south. The breeding male, which had been standing northwest of us about 5 m away, was oblivious to most of this but saw the female getting intent and started toward the hares from the west. They sprang off when he was about 10 m away, and he ran toward them but stopped and looked confused when the hares rushed off. After about 10 sec, he chased the south hare, while the female pursued the north hare over a ridge. A hare and the female wolf appeared a few times on the ridgeline northeast of us, but 5 min later the female was standing while the male was still running down a ravine. About 90 sec later, the male also gave up.

17. July 28, 1992

At 0713 hr, the female wolf jumped a small, grayish leveret and caught it within 30 m. She stood near it for 2 min before eating it and consumed it all.

18. July 28, 1992

At 0803 hr, the female wolf suddenly pounced and picked up a leveret. She then jumped another leveret and chased it out of sight.

19. July 10, 1993

At 1640 hr, the adult male and female wolves were heading west, and the three yearlings were following but lagging 100 m. A yearling jumped a white hare on the north side of the plateau and chased it back and forth for 1 min. We heard no yipping or other sounds. The other two yearlings did not notice the chase for 40 sec, although they were only 50–75 m away. Suddenly, they joined it, and all disappeared into a valley. We then saw that a yearling had caught a white hare, either an adult or a well-developed leveret.

20. July 10, 1993

At 1750 hr, a yearling female caught and ate a nestling, probably a jaeger. Suddenly, a leveret, three-fourths the size of a snowshoe hare, jumped up 2 m away, and the yearling took chase. She usually was within 3–10 m of the hare, and after 450 m caught it. A second yearling had joined the chase when three-fourths over. (This hare had been hiding there while the breeding pair had passed within 6 m, 30 min before.)

21. July 10, 1993 (N. Gibson)

At 1958 hr, a leveret flushed and ran north uphill. A yearling wolf, 200 m away, chased it 500 m more, and the hare circled back. The adult wolf pair had halfheartedly run north uphill, and the female lay on top of the hill, the male 30 m behind. When the hare got closer, the male joined the chase and after 200 m more caught the hare.

The three yearlings mobbed their father, but the yearling that had started the chase ended up with the hare.

22. July 10, 1993

2016 hr, a white adult (?) hare appeared at 40 m away on top of a ridge, and two yearling wolves started toward it. Each stopped on point when the hare sat up. After 20 sec, the hare ran, and they fled after it 150 m behind, but it disappeared. When the wolves searched for it, one jumped two leverets and chased one. As the adult hare and leveret escaped separately, the wolves chased the other leveret, and all five closed in on it. But the hare escaped into a crevice or hole, for the wolves nosed around that spot for several minutes. Two stayed there another 10 min, and one for 5 min more, leaving at 2032 hr.

23. July 11, 1993

At 0540 hr, the five wolves were sleeping when two arose and started chasing a leveret about three-fourths the size of a snowshoe hare. In some tundra crevices 100 m away, it swerved toward where it had come from, and the wolves chasing it could not see it double back. But two other wolves, still at the sleeping spot saw it approach and took chase. The hare headed out of our view, with wolves behind. It popped up again and ran by us 6 m away, with one wolf 2 m behind, and all continued for 250 m where after several zigzags, circles, and so on, one of two wolves caught it. Total chase distance = 500 m. About when the hare came back by the two wolves waiting near the sleeping spot, another hare flushed, and a wolf chased it out of sight, but an adult female wolf later came by us from that area carrying a leveret.

24. July 13, 1993

At 2100 hr, we saw five wolves chasing leverets. Too much was happening to take notes, and we could only see parts of the chases. Once three wolves were each separately chasing three hares. The leverets were in groups of up to six but split up when the wolves chased them. Leverets and wolves were going every which way around us. Three wolves were separately chasing three leverets. They all went out of sight.

25. July 13, 1993

At 2226 hr, the adult male wolf went to the south end of a ridge where he just lay looking around. A leveret went up the hill behind him, and out of our sight, but the wolf suddenly got up and gave chase unseen by us. Probably he did not catch it because, a few minutes later, he showed up along another ridge with a long series of rock piles on it (see below).

26. July 13, 1993

At 2230 hr, a female yearling wolf chased a leveret zigzagging along some rock piles, with the wolf 200 m behind. After losing sight of that hare, the wolf did some ground sniffing and got within 3 m of a leveret when it jumped up. (The same or a different one). It ran 90 m south along the rock piles, and suddenly the adult male wolf popped out from behind a pile, caught the hare, and gave it to the yearling.

27. July 13, 1993

At 2304 hr, the same yearling chased a leveret from around a rock pile for 100 m. The hare doubled back, and the adult male chased it as well, and either one or both caught it. The yearling ended up with it.

28. July 13, 1993

At 2317 hr, another yearling chased a leveret around the rock piles, and the adult male caught it and took it to where the adult female had been. However, she was gone. The male continued guarding the hare for at least 45 min.

29. July 14, 1993

At 0035 hr, a female yearling wolf sniffed around each of many rock piles along a ridge, jumping a leveret from the 21st pile. She chased or followed the leveret 800 m, but the leveret outran her and stopped. Meanwhile another yearling came by, and the hare flushed behind her, so she did not see it. The latter yearling sniffed the ground searching for several minutes in the area 50 m beyond. She then sat on the north end of the ridge watching for several minutes. Apparently, she had seen the hare come to the area when the other yearling chased it.

30. July 14, 1993

At 0211 hr, the adult female wolf was sleeping when a leveret came within 30 m south of her and stopped for 2 min. The wolf looked up and watched it but did not even stand up. Finally the hare headed south 18 m and stopped for 90 sec and then fled. The wolf never arose.

31. July 16, 1993

At 2329 hr, the adult male and a yearling saw two leverets standing 75 m away and although the wolves looked at them, when the leverets ran, the wolves did not run but, instead, continued on.

32. July 17, 1993

At 0140 hr, a leveret ran within 30 m of a yearling male wolf that was resting or sleeping above it on a hill but gave chase. He momentarily gave up when the leveret was 150 m ahead. But then the wolf cut across a gully that the leveret had gone around the rim of. A few minutes later, the yearling appeared chasing a leveret back along the rim into the direction from which it had come. The leveret backtracked, and the adult female wolf, which had been on a hill near the yearling originally, joined the pursuit, shortcut the leveret, and caught it at the bottom of the hill in a gully.

33. July 6, 1994 (Mech 2014)

A 2-yr-old female wolf jumped a leveret (around 20 cm long), chased it for 4 min back and forth, catching it several times with her paws, grasping it with her mouth, shaking it, and dropping it. She could have killed it several times but did not. (As a yearling, she had killed several larger ones quickly.) She then delivered it alive to a pup.

34. July 7, 1994

At 1925 hr, the 2-yr-old female chased a hare for 200–300 m, and the breeding female joined but no success.

35. July 7, 1994

At 2235 hr, the 2-yr-old female headed south uphill and stalked, then watched, two adult hares for 12 min, perfectly still in a crouched position like a cat. The hares appeared to come within 8 m, although for several minutes one stood up still. When the hares started hopping away, the wolf continued watching, and after 30–60 sec, stalked slowly in their direction, but they were out of my sight. The wolf then rushed into a small valley and started zigzagging on a small scale like when chasing a small leveret. The adults were 100 m away. The wolf then came back with a leveret 25 cm long. (Was the wolf biding her time while watching the adult hares in order to catch one of the leverets?)

36. July 16, 1994

At 1445 hr, a 2-yr-old male wolf chased a leveret, and the breeding male and a 2-yr-old female lay on high spots watching. The young male reappeared quickly without the hare, and all three lay on rock piles around 75 m apart watching intently.

37. July 16, 1994

At 1605 hr, the 2-yr-old male suddenly ran straight very intently some 400–500 m right to a leveret (around 1 kg). The leveret was sitting and crouched lower as the wolf approached and did not run until the wolf was 3 m away. It zigzagged a dozen times before the wolf caught it within 1 min.

38. July 16, 1994

The 2-yr-old female poked around various rock piles, apparently searching for hares. At 1612 hr, she chased a 25 cm long leveret but did not get it. She and the breeding male then continued searching rock piles and trying to follow the leveret's trail.

39. July 15, 1996

At 2100 hr, the breeding male ran into a gully and caught a leveret without a chase or with only a very short chase (out of sight) and cached it.

40. July 15, 1996

At 2315 hr, the breeding male chased a leveret for 2 min and 500 m. He even did a complete tumble, head over heels, at high speed but finally caught the leveret and rested for 4 min, then cached it.

41. July 16, 1996

At 0041 hr, the breeding male was clearly hunting hares. He went in and out of gullies and zigzagged, nose low. At 0049 hr, he jumped a leveret and, after 30 sec, caught it and cached it.

42. July 16, 1996

The breeding male chased a leveret at 0114 hr and caught it out of sight, but I saw him with it at 0116 hr.

43. July 16, 1996

At 0136 hr, the breeding male caught a leveret after a short chase.

44. July 16, 1996

At 0146–49 hr, the breeding male chased a leveret back and forth, starting with the leveret about 100 m away, but had almost caught it when he went into a large valley out of sight.

45. July 17, 1996

At 0420 hr, the breeding female started sniffing the air toward the west. The wind was from the northwest. The wolf then headed east excitedly. She hurried alertly east for 200 m and then rushed another 100 m east and chased two leverets for 1 min. The last half of the chase was out of sight, but she appeared with one of the leverets after 1.5 min. The wolf seemed to have sensed the hare by smelling in the opposite direction. Did the northwest wind switch in the valley, and the wolf knew it?

46. July 22, 1996

At 0705 hr, the breeding male and female chased a leveret, but the wolves gave up.

47. July 22, 1996

At 0720 hr, the breeding male chased a leveret for 30–45 sec and missed it. He and the breeding female then sat on a hillock and kept looking around, both watching for leverets. At 0731 hr, the male led the female down the hill 30 m and pounced on a leveret hiding in some hummocks.

48. July 22, 1996

When the breeding male wolf was within 2–3 m of a hare, the hare flushed, and the wolf chased it about 500 m and then gave up.

49. July 22, 1996

At 0051 hr, the breeding male chased a leveret as his mate watched from on top of a hill. As the male chased the hare around the south side of a large hillock; the female headed around the north side. The hare came around the north side, and the female almost caught it but missed. Both wolves chased it for 2 min to the bottom of the hill, where it hid. After trying to get it out for about 5 min, they gave up.

50. July 22, 1996

At 0141 hr, the breeding pair of wolves were in the vicinity of an old musk ox carcass and hunted gullies. They each watched the other, although the female tended to take her cues from the male. They often stopped on high points and scanned the area below for 30–120 sec. At one point, the female stalked a group of more than eight leverets and suddenly ran toward them. I lost sight of the wolves, but suddenly the male came charging in the opposite direction chasing a "cloud" of 16 light-gray leverets. They all disappeared over the bank, and individual hares were running all over some flats and adjoining hills, but I lost track of the wolves (0200 hr). I caught intermittent views of both wolves until 0235 hr, and they did not act like they had caught any of the leverets. That is, they did not seem to be eating or carrying or caching any leverets and were not out of sight long enough for them to have done any of these things, but I cannot be certain.

51. July 25, 1996 (Mech 1997)

The 11–13-yr-old male wolf slept; at 0535 hr, he walked 100 m and suddenly veered toward a crouching leveret about 10 m away and more or less upwind of him. He walked by the hare, passing 3–7 m by it, got 7 m beyond, turned, and headed toward it. The hare then jumped up, and the chase began. The wolf chased the hare for 6–7 min and almost caught it several times, but the hare's ability to make quick turns helped it elude the wolf since the wolf could not turn so sharply. The chase went back and forth, up and down gently sloping hills covering a distance with a maximum radius of 300 m. At times, the hare was as far as 30 m ahead of the wolf. Finally, at 0544 hr, the hare seemed to tire and slow down, and the wolf pounced on it. He carried the hare off and cached it. It weighed 1.45 kg.

52. July 28, 1996

The breeding female was leading, hunting hares. Her mate generally lagged by 200–1000 m. 0952 hr—Both chased a leveret for 30 sec and gave up.

53. July 28, 1996

The female chased a group of four leverets, and especially one of them that cut downhill, but after about 1 min, she gave up as the hares easily outran her.

As apparent from the above observations, wolves must use many different strategies for hunting arctic hares because of the great differences in size, life style, and physical abilities of hares from birth to maturity. When hares are very young, wolves seem to connect the behavior of nursing females with the presence of vulnerable leverets (accounts 5, 35). As the cryptically colored leverets grow and spend their nonfeeding time crouching still, wolves sometimes watch from vantage points and seem to be able to spot them as far as 400–500 m away, perhaps seeing a twitch or other movement. Then they merely approach and chase them (account 37) or pick them up (account 47). Other times, wolves accidentally flush the leveret when walking by, then chase it until they overrun it or exhaust it. However, this vulnerable period for the hares only seems to last a few weeks or less. My field notes indicate the following: "It is notable that the [breeding] male missed 3 leverets today [July 22, 1996] and caught only one compared with 6 successes last week. It appears they are getting large enough to begin outrunning the wolves."

If the leverets join a nursery group, the relative rarity of such groups (as opposed to the greater number of spread-out individuals), the confusion, and the sheer number of associates reduces the vulnerability of any single leveret. Based on accounts 15, 22, 24, and 50, the wolves appeared less efficient at catching leverets when the leverets were in groups or when there were large numbers of them because of the confusion of too many running every which way. The wolves did catch several and seemed to do better after they had split the group up. Furthermore, if several wolf-pack members hunt leverets together, then the leverets are vulnerable to ambushing (deliberate or accidental) by one wolf while the others chase them (accounts 26–28, 32). Presumably this technique would work with adult hares as well, but too few observations of wolves hunting adult hares have been made to determine this possibility.

Wolves were successful at catching arctic hare leverets 60% of the time in July, with single wolves succeeding in 16 of 26 attempts and multiple wolves succeeding in nine of 16 attempts. Anecdotally, it appeared that success rates dropped considerably as leverets grew, but the sample of observations was too small to allow for statistical testing.

Following are some other wolf hare-hunting generalizations:

1. Much hare hunting is opportunistic—the wolves chase hares when they jump them while traveling or when the hares are resting or sleeping and the wolves spot them ("Arctic hare [*Lepus arcticus*]," BBC Natu-

ral History Unit, http://www.arkive.org/arctic-hare /lepus-arcticus/video-11.html). However, when wolves are resting or sleeping, they seem to be able to detect adult hares readily. The wolves must always be "keeping an eye out" even when seemingly asleep.

2. Wolves can see hares a long distance—probably as far as humans—and seem to be interested in the hares when up to 400–500 m away. That is, when wolves see leverets as far away as this, they head to them.

3. The adult male wolf seems to lie in wait to ambush hares, especially when the yearlings are chasing them.

4. Yearlings seem faster at chasing hares than do the adults. (Mech wrote this in field notes in 1993, and MacNulty et al. [2009a] tended to confirm it with wolves chasing elk).

5. Wolves can and do chase hares long distances, sometimes more than 800 m.

6. When in an area with many leverets, wolves sometimes split up and lie not more than 200 m apart and watch for hares.

7. When hunting hares, the breeding pair sometimes travel about 25 m apart, probably to increase chances both of finding a leveret and catching one.

Because the manner in which wolves eat hares has not been described, we include the following excerpt from Mech's July 19, 1994, field notes:

We watched [2-yr-old female] Explorer lying near [2-yr-old male] Grayback II while he ate an adult hare. He ate insides first leaving stomach and guts, then down to hind legs and up to head, kind of peeling the meat out of the skin and eventually chewed through the skin holding front and back ends together, leaving 2 pieces. He looked like he was going to take the hind quarters to the den, but when Explorer [2-yr-old female] came closer to him, he started eating the hind legs. He let Explorer poke through the guts and pick what she could but she left some of them. She ate the hind foot Grayback II [2-yr-old male] left. (She had also eaten the hind foot the breeding male left.)

Snowshoe Hares

Snowshoe hares inhabit much of the wolf's range in Canada and the northern United States. They weigh 1.0–1.8 kg, so they do not provide much of a meal for a wolf.

Furthermore, their numbers wax and wane. Thus, it is not surprising that wolves generally do not expend much time or energy seeking them, and in some cases do not even chase them when flushed (Mech 1966a; Peterson 1977). However, at other times, snowshoe hares become plentiful and can supply at least supplementary food (Stanwell-Fletcher and Stanwell-Fletcher 1942). Whether wolves chase snowshoe hares or not probably depends a great deal on how hungry a wolf is, how easy it seems to catch the animal, and so forth. In any case, wolves do sometimes chase and kill them. Following are a few such accounts.

1. January 18, 1966—Superior National Forest, Minnesota (Mech 1966b)

At 1440 hr, one of seven wolves chased a snowshoe hare, which ran under a blowdown; the seven wolves surrounded the blowdown and stood with tails wagging. There was no indication that they caught the hare. They nosed around the vicinity of the blowdown for a few minutes before continuing on.

2. March 28, 1977—Superior National Forest, Minnesota (J. Renneberg)

At 0825 hr, I watched adult male Wolf 5132 and another wolf chasing two snowshoe hares. Just a short chase and gave up quickly.

3. September 27, 1991, Denali Park, Alaska (Mech et al. 1998)

Tom Meier recorded the almost absurd chase by 13 East Fork wolves of snowshoe hares in the Wyoming Hills area northwest of Toklat and saw them catch one and miss one; the remains of two hare kill sites were also visible in the immediate area.

Salmon

In some areas of Alaska and British Columbia, wolves find that one of their least risky prey is salmon (Jackson 1942; Stanwell-Fletcher 1942; Darimont and Reimchen 2002). No doubt in areas where and when other fish are plentiful, wolves take them too. The only written account of how wolves hunt salmon comes from Young and Goldman (1944), citing Jackson (1942, 29): "Once a salmon splashed in the shallow stream and two of them [wolves] raced after it. The first wolf to reach it grabbed the salmon by the back and brought it up onto the bank. There he lay down with the fish between his paws and ate off its head." One can watch wolves in the act in two YouTube videos ("Alaskan Wolf Catching a King Salmon," uploaded August 3, 2009, http://www.youtube.com/watch?v=IXiGvoKdCIM; and "Wolf Catching Salmon at Brooks Falls," filmed by Peter Hamel, uploaded November 13, 2008, http://www.youtube.com/watch?v=WyFjxm_aSEU&feature=related).

Waterfowl

Like salmon, flightless (molting) waterfowl sometimes also can provide a bonanza for wolves, as wolves merely run them down (D. W. Smith, pers. comm.). The following describes signs of wolves having killed many such ducks (Hagar 1968).

> Our attention was attracted to a network of beaten trails through the grass, and to places here and there where feathers and bits of bone were scattered about on the flattened vegetation over a diameter of 5–6 feet [1.5–1.7 m]. We presently found that these trails and trampled spots conformed roughly to a pattern: virtually every pond-hole in the meadow was circled by a path some 10–12 feet [3–4 m] back from the edge of the water, the beaten-down openings were strung along these circular trails, and the remaining trails for the most part ran from one pond to another. Closer examination of the openings showed that the feathers were those of ducks, many of them being the partly chewed ends of primary feathers still in sheath, and that some bits of bone could also be identified as coming from ducks, although in no case was any large bone left whole . . . I felt reasonably sure that the wolves were trotting systematically around each pond, nosing out moulting ducks from the grass and when one was caught, lying down at once to eat it.

Similarly, wolves sometimes pilfer nests of geese and other ground-nesting birds, as well as taking the adults when possible.

1. May 21, 1992—Copper River Delta, Prince William Sound, Alaska (Stephenson and Van Ballenberghe 1995b)

We made this observation while circling in a Piper PA 18-150 Supercub fixed-wing aircraft. At 2110 hr, a radio-collared, adult female wolf flushed an adult dusky Canada goose from a nest containing at least four eggs and consumed the eggs. The wolf then traveled 300 m, flushed a second goose from a nest 50 m away, and, after failing to locate the nest during a quick search, departed.

2. May–July 1992, 1994, 2000–2004—Queen Maud Gulf Bird Sanctuary, Nunavut, Canada (Wiebe et al. 2009)

We saw wolves traveling among nesting geese 27 times. All geese less than 20 m from the wolf ($n = 20$) or wolves ($n = 7$) left and settled 50–100 m away. Geese often walked, flew, or swam 100–200 m behind wolves. We saw wolves rushing at geese only three times (during one of which a wolf rushed at geese eight times). All rushes covered only a few meters and were directed at geese late leaving their nests. Once a wolf fed on two eggs after nudging them out of a nest down a 2 m high bank (presumably breaking them). Another time, a wolf ate eggs from three different nests during the same observation. Wolves ignored more than 1,000 geese and their eggs during our observations.

3. May 28, 1992—Copper River Delta, Prince William Sound, Alaska (Stephenson and Ballenberghe 1995b).

During 1115–30 hr, a radio-collared, adult female wolf consumed the eggs of three goose nests. The nests were located by sequentially flushing the adults as the wolf searched a broad sedge meadow. After consuming the third clutch, the wolf behaved in an animated manner and chased its tail.

4. August 8, 1993—Copper River Delta, Prince William Sound, Alaska (Stephenson and Van Ballenberghe 1995b)

Two radio-collared adult wolves, male and female, were located near a high concentration of molting dusky Canada geese, 1920–45 hr. The wolves were traveling about 50 m apart with the female leading toward a large pond to the northwest occupied by at least 50 geese.

When within 100 m of the pond, the wolves separated; the female crouched and moved a short distance through a low willow-graminoid community. Some geese on the perimeter of the pond apparently detected her because there was a sudden large flush into the pond. The female then ran to the pond, jumped in, and began to chase geese.

The male circled to the opposite side and entered the water when the geese were pushed there. By this time, about half the geese had flown to an adjacent pond (75 m northwest), and the wolves were chasing the remainder. Most of the geese could fly, but apparently not well enough to reach the other pond easily because they were still molting. The shallow water in the pond enabled the wolves to leap (as opposed to swim) and thereby exhaust the geese.

After 5 min of pursuit, the male captured a goose, carried it to shore, and returned to the pond. During the next 6 min, working as a team, each caught a goose. Most geese had flown to the larger, adjacent pond 75 m northwest, and the female crossed the narrow strip between the ponds and began chasing geese in the shallow end of this pond. The geese moved to deeper water, however, and the female had to swim and quickly fell behind.

Small Mammals

Wolves do eat mice, lemmings, and other small mammals, but contrary to the popular, fictional account *Never Cry Wolf* by Farley Mowat (1963), nowhere is it documented that wolves subsist on them for long periods. It is true that earlier writers did state that during years of lemming or mouse peak populations, wolves, as well as "all predators, including birds, mammals, and even fishes, as trout, joined in living almost exclusively on mice" (Cabot 1920, 292). This claim was not documented, and similar claims have not been made since then. Nevertheless, like any similar tidbits such as eggs, nestlings, berries, and even garbage, mice often afford wolves another tasty morsel. Only Murie (1944) has described wolves hunting mice, although hundreds of laypeople certainly have watched them do so in Yellowstone Park.

Murie (1944, 57) also related an interesting anecdote related to wolves eating mice: "One evening in September I tossed in quick succession to the tame wolf pup, which had a surplus of dried salmon and scraps available and was not especially hungry, 19 mice and two shrews. These were caught in the air and swallowed so rapidly that each one hardly stopped in the mouth. There was no evidence of satiety after the last mouse was swallowed."

1. May 3, 1940—Denali Park, Alaska (Murie 1944)

An adult wolf hunting a mouse in the snow pounced with legs stiff, much in the manner of a coyote, to break through the crust. After the pounce, he dug in the snow, but I could not be sure that the mouse was captured.

2. September 10, 1940—Denali Park, Alaska (Murie 1944)

An adult wolf was hunting mice for an hour on a flat and made about 17 pounces. It then traveled briskly, turning aside occasionally to catch a mouse that it heard or smelled along the way. Usually in pouncing on a mouse it did not hop up in the air and drop on it, but pounced in a more nearly direct horizontal line, much as a dog would do. Once, it went up in the air more than usual and so vigorously that on landing it went off balance and its hindquarters swung to one side. Often it made a short run of five or six jumps before pouncing. When looking for a mouse—one for which it had pounced and not captured—it wagged its tail, the rate of tail wagging increasing with the rise of its excitement, which in turn was dependent on the nearness of the mouse.

3. September 17, 1940—Denali Park, Alaska (Murie 1944)

A black pup hunted mice by itself 800 m from the other wolves for 2 hr. One pup played with one it had captured, tossing it in the air a few times and catching it. Apparently the pup was not hungry. At 1315 hr, after a rest of 3 hr, that pup went off hunting by itself. The usual method of pouncing was noted, but sometimes there was hardly any jump, and only the forepaws would strike forward to press the mouse to the ground.

4. September 24, 1940—Denali Park, Alaska (Murie 1944)

Two pups hunted diligently for mice 800 m away 0800–1200 hr. In the afternoon one of the pups was again seen hunting mice. At this time, when the main caribou herds were elsewhere and only stragglers were available, the mice seemed to form an important supplement to the food supply, especially for the pups but also for the adults.

Conclusion

Many of the miscellaneous prey wolves take are killed either incidentally to their other hunting and traveling or under limited, specialized circumstances. Examples of the latter are spawning salmon, molting waterfowl, and the sudden flush of arctic hare leverets. In these cases, wolves are merely taking advantage of local, temporary opportunities. In some of these situations, these kinds of nonungulate prey can help local wolves through periods when ungulate prey might be less available. In the case of arctic hares, evidence suggests that some arctic wolves might depend heavily on them during summer (Tener 1954a, Mech 2007c).

In many situations in which wolves prey on smaller creatures, no specific hunting behavior or strategy is necessary other than to grab the prey or pounce on it. Even so, wolves may well end up with individuals that tend to possess poorer senses or are smaller or poorly developed (Wirsing 2003).

9

Conclusion

IN THE PRECEDING chapters, we have examined the wolf as a killing machine, the "beast of waste and desolation," in Teddy Roosevelt's words. Bearing long, sharp fangs, howling in the night, and traveling in packs, wolves gave some early observers the impression that wolves could kill just about anything at will. J. R. Mead (1899, 280), in his "Some Natural History Notes of 1899," for example, stated that wolves "could kill the strongest bull [bison], and did, whenever appetite and circumstances made it most convenient."

Detailed, painstaking, time-consuming observations, however, were not the forte of such early observers. Rather, they got to see and record the superficial and obvious, like remains of wolf-killed prey scattered across the plains or strewn on blood-soaked snow. Similar types of observations are still made by many laypeople, usually with the same resulting conclusions. "The deer [moose, sheep, caribou, bison etc.] was perfectly healthy," many exclaim, only having surveyed a few scattered bones, and taking a few pictures. "Wolves kill for sport, often bringing an animal down, mauling it, ripping the gut open, then leaving the animal to die a slow, torturous death," reads a recent website posting (Lynn Stuter, "The Truth about the Wolves," pt. 2, *Lynn Stuter Commentary* (blog), February 9, 2010, http://www.newswithviews.com/Stuter/stuter176.htm).

Modern scientists, however, greatly differ from old-time naturalists and current laypeople in their approaches to gathering information and making conclusions about wolves and their behavior. In that respect, they have far more advantages for determining reality. Not only are they trained to make careful, detailed observations over long periods and large areas, but they also possess the technology—such as aircraft, radio tracking, blood analyses, and genetic assessments—necessary for a scientific determination of actuality. In addition, instead of enjoying the constant reinforcement of their conclusions by like-minded friends, scientists and their observations are subjected to the constant critical scrutiny of other scientists through the peer-review and publication process, all with the ultimate goal of determining the actual reality rather than the positing of hasty conclusions that may well be wrong.

The end result, then, is that by examining the carefully collected data, scrutinizing the many observations from the preceding chapters, and synthesizing the information from all, we can gain a closer view of reality. The reality of wolf hunting behavior is that it is far more complex than the notion that the wolf is a killing machine that can slaughter anything it wants anytime (www.press.uchicago.edu/sites/wolves, video 17). Under the right circumstances, the wolf can and does efficiently catch, subdue, and kill prey much larger than itself, just as, under the right circumstances, the large prey can survive in the face of numerous attempts by wolves (and other predators) to kill them. Once they survive for 1–2 yr, bison, elk, moose, deer, and other primary prey of the wolf commonly reach 10, 15, or 20 yr of life or more. This means that, for all those years, these creatures thwart the wolves' attempts to kill them. One study estimated that, during summer, at least one wolf visited an average deer's home range every 3–5 d (Demma et al. 2007). Another study found that during winter each elk in one Yellowstone herd was visited by wolves within a kilometer every 9 d on average (Middleton et al. 2013). If this is typical, think of how invulnerable these prey animals must be.

As already mentioned, at least some prey species in some areas do not even try to avoid wolves (Soper 1941; Banfield 1954; Fuller 1960; Mao et al. 2005; Kittle et al. 2008), contrary to the "ecology of fear" idea (Brown et al.

1999; Ripple and Beschta 2004; Berger 2008). Mech has even watched individual musk oxen approach active wolf dens, and deer with fawns lived regularly within 1.8 km of active wolf dens (Nelson and Mech 2000).

The above observations tend to contravene much of the current thinking reflected in recent scientific literature on predator-prey interaction theory. According to MacNulty (2002), predators are usually considered "fierce" (Brown et al. 1999) because they elicit a fear response from prey that involves the prey making trade-offs between food and safety to survive and reproduce (Krebs 1980; Newman and Caraco 1987). The fitness consequences related to this trade-off result in the evolution of adaptive foraging strategies. Thus, most foraging studies focus on the behavior of an assumed timid prey pursued by a fierce predator and assume that predators hunt prey with no risk of prey-caused injury (Sih 1980; Caraco 1981; Dill 1987).

However, MacNulty (2002) and Mukherjee and Heithaus (2013) have suggested that wolves (and other predators) hunting dangerous prey respond to the risk of injury, so they also must make trade-offs between food and safety. In this case, safety represents the critical need to avoid being killed or injured by belligerent prey. Because predators cannot simultaneously maximize food intake and minimize injury risk, conflict arises in deciding whether to attack or avoid dangerous prey. Wolves appear to have resolved this conflict by evolving a tendency to take vulnerable prey. As a result, wolves can acquire food while avoiding or minimizing the risk of injury (Mech 1970).

Since the fitness consequences of prey-caused injury are severe, the risk of injury is likely a strong selective force over evolutionary time on the foraging behavior of predators that rely on dangerous prey for food. Further insight to foraging behavior is possible by moving beyond the traditional view of a timid prey pursued by a fierce predator (Lima and Dill 1990; Brown et al. 1999) and considering more behaviorally sophisticated systems such as those we have described in the preceding chapters where prey are not invariably timid, and wolves are not fierce enough to kill them at will.

Sooner or later, most wolves are successful at finding a creature they can catch and kill without too much danger to themselves. The traditional view of this interaction is that all wolves need to do to find, catch, and kill their quarry is to travel and search enough until they locate a vulnerable individual (Mech 1970; Mech and Peterson 2003). And there certainly is sufficient evidence that most of the prey that wolves catch show some kind of vulnerability (below). However, there is more to the total picture.

A newer approach to analyzing wolf interactions with prey examines the possible role of landscape features, for conceivably landscape could affect the nature of each of the stages of the hunt. The "terrain trap" discussed in chapter 4 is a dramatic example. The first study of landscape features assessed only the landscape where wolf kills were found, not necessarily the prey's original location or where the encounter or chase took place, which are other critical aspects of the wolf-prey interaction (Kunkel and Pletscher 2000). Next came an analysis of actual travel routes of hunting wolves. That study found that wolves in a deer, elk, and moose system tended to travel primarily in deer winter ranges even though the wolves killed elk and moose disproportionately more (Kunkel et al. 2004). The explanation was that deer were more often encountered or easier to find, so wolves chose to travel in their habitat.

More recently, the habitat of encounter location (between wolves and elk) was studied along with that of the kill locations. That study learned that "the odds of elk being encountered by wolves was 1.3 times higher in pine forest and 4.1 times less in grasslands than other habitats. The relative odds of being killed in pine forests, given an encounter, increased by 1.2. Other habitats, such as grasslands, afforded elk reduced odds (4.1 times less) of being encountered and subsequently killed (1.4 times less) by wolves" (Hebblewhite et al. 2005, 101). Unfortunately that study did not assess the age, sex, or condition of the 119 wolf-killed elk found to determine whether, for example, it was the more vulnerable elk that inhabited pine forests where odds of being killed by wolves were higher. Perhaps older elk, which are more vulnerable to wolves (Mech et al. 2001, Smith and Bangs 2009), tend to inhabit pine forests with greater frequency, much as senescent moose inhabit different habitat from that of prime-age moose (Montgomery et al. 2013). In any case, this type of more refined examination of various factors involved during wolf hunting of prey (see also Atwood et al. 2009; Kauffman et al. 2007) is sure to add considerably to our knowledge of the wolf hunt.

For now, it is clear that generally the most important factor driving wolf-prey interactions is selection of prey, which is an important means by which wolves can kill prey while minimizing the risk of injury or death. Prey selection involves a variety of factors, including selection of specific prey species in multiprey systems, age, sex, and condition of prey killed, and such nebulous factors as blood values, behavior, sensory abilities, and multigenerational effects on prey vulnerability. Mech (1970) devoted a whole chapter of his book *The Wolf* to selection of prey, and Mech and Peterson (2003) updated the subject.

Suffice it to say here, in summary, that it is well documented in the above two publications and elsewhere that wolves generally kill calves, fawns, and older members of prey populations along with individuals that are diseased, disabled, or in poor condition or that have various abnormalities. These types of individuals are physically less able to withstand long and persistent attacks like more healthy animals can. These prey are also safer to kill, which minimizes their net cost. Paquet's (1989) finding that the length of a wolf's chase correlated well with the quarry's fat stores is elegant testimony to that. Furthermore, evidence indicates that it is not only because the wolf hunting process ensures that these types of individuals tend to fall prey to wolves more frequently but also because, at least at times, wolves tend to seek out or target such vulnerable individuals. Sokolov et al. (1990) demonstrated this possibility using borzoi dogs, but Landis (www.press.uchicago.edu/sites/wolves, video 7 [1998]) filmed wolves in Yellowstone National Park actually targeting a limping elk with arthritis out of a larger herd and killing it.

It certainly is not unreasonable that a creature with the wolf's intelligence that is hunting prey daily could learn to discern prey that are slower, weaker, debilitated, or in other ways abnormal. Even human hunters in at least some cases find that quarry with more parasites are more vulnerable (Rau and Caron 1979). The Landis video (www.press.uchicago.edu/sites/wolves, video 7 [1998]) portrays two black, yearling wolves slowly chasing the elk herd and apparently scanning it for several minutes before suddenly increasing their speed, on targeting a limping elk, from 14 km/hr to 45 km/hr (MacNulty 2002). The stride frequency of one wolf increased from 2.1 to 2.4 strides/second, and the stride frequency of the second wolf increased from 1.9 to 2.6 strides/second. Assuming a constant stride length, these figures reflect a 14% and 33% increase in speed, respectively. Using a larger sample ($n = 18$ wolves), MacNulty et al. 2007 found that stride frequency increased from 2.06 to 2.43 stride/second once wolves targeted a specific elk.

In the Landis video (www.press.uchicago.edu/sites/wolves, video 7 [1998]), that limping, arthritic elk is even apparent to humans viewing the video, but very probably well-practiced wolves could detect much more subtle vulnerabilities. One such vulnerability bears further discussion because it constitutes an excellent example of the type of subtle abnormalities that Mech (1970, 248) cautioned about: "We probably could not recognize most of the circumstantial, behavioral, or physical conditions that might have made prey vulnerable." That vulnerability has been referred to as "the grandmother effect." Long known in nutritional experiments with rats (Zamenhof et al. 1971; Bresler et al. 1975; Chandra 1975), this effect has since been documented in white-tailed deer (Mech et al. 1991) and more recently in humans (Kaati et al. 2007). Basically, the studies show that animals that are poorly nourished (F_0) while carrying fetuses produce grandoffspring (F_2 generation) that are physically inferior in several ways regardless of how well their mothers (the F_1 generation) are nourished. With deer, significantly more second-generation individuals from 0.8 to 2.0 yr of age whose grandmothers were poorly nourished were killed by wolves than those whose grandmothers were well nourished, regardless of how well nourished their mothers were (Mech et al. 1991). Although the mechanisms of this transgenerational relationship have long remained unknown, it now appears that epigenetic mechanisms might be involved (Kaati et al. 2007).

An investigator examining the remains of a wolf-killed deer might conclude that it was in excellent health and condition, not knowing anything about its grandmother's nutrition. Similarly, substandard senses of smell, sound, or sight (Mech 1970), having fewer brain cells (Zamenhof et al. 1971), or any number of other conditions might predispose an animal to wolf predation. One example is the case of prime-aged moose killed by wolves on Isle Royale that when examined closely tended to be unusually small (Peterson 1977). Thus the common claim, even by some biologists, that a wolf-killed animal "was perfectly healthy" just because it was of prime age and fatty and appeared to be in good condition cannot be supported.

The question of whether wolves tend to favor hunting certain species over others in multiprey systems has been the subject of several studies, with variable results. At first thought, it seems logical that some species would be preferred over others, deer over moose, for example. However, in a deer/moose/elk system where there were an estimated 10 times the number of deer than either moose or elk, wolves took significantly more elk and moose than deer (Kunkel et al. 1999, 2004). Mech and Peterson (2003) reviewed the literature on the subject, and after discussing several considerations concluded the following:

> While elements of learning, tradition, and actual preference may be involved in apparent prey species preferences, the most likely explanation for these patterns involves a combination of capture efficiency and profitability relative to risk, which boils down to prey vulnerability. In other words, we believe that as wolves circulate around their territories and encounter and test prey under constantly changing conditions, they gain information about the relative vulnerability of various types of prey to hunting (including finding, catching, and killing). Through trial and error they end up with whatever prey they can capture. Thus as conditions change, the wolves' prey changes in species, age, sex, and condition.

Because less-fit individuals vulnerable to wolves necessarily comprise only a small proportion of most prey populations, wolves must be ready to catch and kill such animals whenever they encounter them. However, occasionally circumstances change such that a higher proportion of vulnerable individuals are suddenly available. Such circumstances can be weather conditions, such as especially deep snows or prolonged drought, or they can just be the sudden birth pulse of a caribou herd. If wolves find groups of vulnerable prey due to such circumstances, they usually follow their instincts and kill as many such prey as they can. Migrating salmon are a good example of this practice. Wolves in Alaska and the west coast of Canada commonly catch large numbers of salmon, and often "high grade" them, eating only the

heads (Jackson 1942; Darimont and Reimchen 2002). Similarly, "when caribou are abundant, wolves often kill in excess of their immediate needs" (Banfield 1954, 49). Several other instances have been documented of wolves killing large numbers of caribou calves, adult caribou, reindeer, elk, and white-tailed deer and not immediately eating all of them (Mech et al. 1971, Miller et al. 1985; Del-Giudice 1998; Mech et al. 1998, Mech et al. 2001; Mech and Peterson 2003).

Originally dubbed "surplus killing" (Kruuk 1972), this phenomenon has also been called "excessive killing" (Carbyn 1983). A fine distinction between the terms has been proposed (Miller et al. 1985), but it is arguable whether this distinction has any relevance. In both cases, wolves kill more than they can immediately consume. However, if sufficient time passes without scavengers finishing off the carcasses, wolves do return and feed on what researchers originally regarded as surplus or excessive (Mech et al. 1998, 2001). Banfield (1954, 49) perceived this behavior more than a half-century ago when he wrote that "what may appear to be excessive killing by wolves seldom goes unutilized in the long run." The fact is that wolves are programmed to kill and eat whenever they can because generally it requires considerable time, energy, and risk most of the time to do so.

In the introduction, we discussed the thoughts of various biologists regarding wolves' use of strategy and cooperation, but the evidence for this use in the foregoing observations of wolves hunting is not strong, although some is tantalizing. In certain respects, conclusions about this subject depend on definitions. When wolves travel as a group, each in line behind the other, is this coordination? Cooperation? What about when the lead wolf stops alertly viewing prey, and the others do the same? When several spread out and conceal themselves watching prey and then all charge the prey at once on a broad front (Clark 1971; Mech 2007b)? Then there is the question of interpretation of apparently patterned hunting behavior and the finding that robots using only simple rules can display such behavior (Muro et al. 2011). Because of all these considerations, we remain open-minded about the question of how much deliberate strategy wolves use during hunting and how complex that strategy might be.

In any case, after all the investment wolves must make in locating, chasing, and catching their quarry, they still must stop it and subdue it to the point where they can begin consuming it, the "capture" phase of MacNulty et al. (2007). As Durward Allen (1979, 128) wrote in his inimitable way, "about all most meat eaters require of their prey is that it stand still." Sometime during this process, the quarry dies. Wolves have no trouble dispatching smaller prey like hares, of course, merely by grabbing them, but larger prey require considerable attention and effort. Most often, larger prey that wolves attack are fleeing from them, so it is only natural for wolves to leap at their hind end, grabbing wherever they can, usually the rump or a hind leg. A hind leg or rump grip tends to slow the quarry, and other wolves can then attack the front end. In Yellowstone National Park, wolves grabbed prey <200 kg by the neck only, prey 200–270 kg by the hind end and neck, and prey >270 kg by the hind end (MacNulty 2002). Contrary to popular thinking, there is no good evidence that wolves hamstring their prey (Mech 1970; Buskirk and Gipson 1978).

With prey the size of deer, elk and caribou, wolves usually tear at their throats (www.press.uchicago.edu/sites/wolves, video 18). There, the vital windpipe, jugular vein, and carotid artery are all exposed to the wolves' fangs. In moose, wolves often attack the nose and, in musk oxen and wild boar, the head. A nose hold or head hold can distract the quarry while other wolves work on the back end (Mech 1966a). If wolves can open up the anal area and pull a gut out, the result would be considerable internal bleeding. Once the quarry drops and stays down, the wolves begin feeding even if the animal is not quite dead (Carbyn et al. 1993).

And so it goes, day after day, as wolves continue their rounds, ever searching for more vulnerable prey animals—chasing, missing, trying again and again, and eventually connecting. The net result of all this sifting and selecting of prey over eons is that prey gradually get faster, smarter, and more alert. "Thus, the young of every kind—chicks, nestlings, fawns, calves, cubs, pups—are a part of the communal food supply and their numbers will be drastically reduced in the natal year. Those that survive to produce another generation are the select few

that continue the slow evolutionary course toward a more competitive level of adaptation" (Allen 1979, 113).

One might think that natural selection, which gradually changes traits and fosters evolution, takes so long that such changes would be imperceptible. However, in the past few decades, many studies of the role of predation in natural selection have been published (Kingsolver et al. 2001). For example, Wirsing (2003, 315) found that smaller juvenile snowshoe hares were disproportionately vulnerable to predation, "perhaps because this measure [size] is positively correlated with escape speed."

Similarly, Barber-Meyer and Mech (2008, 8) reviewed studies of predation on young ungulates, and they found "that successful predation on juveniles is more selective than is often assumed. Because we are unable to control for (or in some cases even measure) the myriad of other possible vulnerabilities such as differences in sensory abilities, intelligence, hiding abilities, tendency to travel, etc., finding selective predation based on the relatively few differences we can measure is noteworthy and points to the significant role that predation on juveniles has in the natural selection of ungulates."

The hunting behavior of wolves, in all the interesting variations we have described, not only is integral to the wolf's survival but also contributes to the overall evolutionary processes functioning throughout the natural world. These processes have helped shape the fleetness of the deer, the alertness of the elk, and the aggressiveness of the moose. To keep up with these defenses, the wolf, too, must continue its long evolution of ways to overcome them, ever searching for opportunities and cashing in on vulnerabilities.

This age-old evolutionary arms race will continue its fascinating progress as long as we humans allow enough natural space for the wolf and its prey to function. We hope this book contributes significantly to an understanding of wolf hunting behavior in a way that fosters the public will to ensure such space will always be available.

Appendix

List of Scientific Names of Birds and Mammals Mentioned

SPECIES	SCIENTIFIC NAME
African hunting dog	*Lycaon pictus*
American lion	*Panthera atrox*
Arctic ground squirrel	*Urocitellus parryii*
Arctic hares	*Lepus americanus*
Beaver	*Castor canadensis*
Bison	*Bison bison*
Camel	*Camel* sp.
Caribou	*Rangifer tarandus*
Chamois	*Rupicapra rupicapra*
Deer, mule	*Odocoileus hemionus*
Deer, red	*Cervus elaphus*
Deer, white-tailed	*Odocoileus virginianus*
Dog	*Canis lupus familiaris*
Elk	*Cervus elaphus*
Grizzly bear	*Ursus arctos*
Horse (wild)	*Equus caballus*
Hyena	*Hyaenidae*
Ibex	*Capra ibex*
Jaeger, long-tailed	*Stercorarius longicaudus*
Jackrabbit	*Lepus* sp.
Mammoth	*Mammuthus* sp.

SPECIES	SCIENTIFIC NAME
Mastodon	*Mammut* sp.
Moose	*Alces alces*
Mountain goat	*Oreamnus americanus*
Musk ox	*Ovibus moschatus*
Pronghorn antelope	*Anlilocapra americana*
Saber-toothed cat	*Smilodon* sp.
Salmon	*Oncorhyncus* sp.
Seal, ringed	*Pusa hispida*
Sheep, bighorn	*Ovis canadensis*
Sheep, Dall	*Ovis dalli dalli*
Sheep, Stone's	*Ovis dalli stonei*
Short-faced bear	*Arctodus* sp.
Snowshoe hare	*Lepus americanus*
Squirrel, red	*Tamiasciurus hudsonicus*
Tahr	*Hemitragus jemlahicus*
Takin	*Budorcas taxicolor*
Weasel, short-tailed	*Mustela erminea*
Wild boar	*Sus scrofa*
Wolf	*Canis lupus*
Wolf, dire	*Canis dirus*

Literature Cited

Adams, L. G., and B. W. Dale. 1998. Timing and synchrony of parturition in Alaskan caribou. *J. Mammal.* 79:287–94.

Adams, L. G, B. W. Dale, and L. D. Mech. 1995. Wolf predation on caribou calves in Denali National Park, Alaska. Pp. 245–60 in L. N. Carbyn, S. H. Fritts, and D. R. Seip, eds., *Ecology and conservation of wolves in a changing world.* Edmonton: Canadian Circumpolar Institute.

Allen, D. L. 1979. How wolves kill. *Nat. Hist.* 88:46–51.

Altmann, M. 1958. Social integration of the moose calf. *Anim. Behav.* 6:155–59.

———. 1963. Naturalistic studies of maternal care in moose and elk. Pp 233–53 in H. L. Rheingold, ed., *Maternal behavior of mammals.* New York: J. Wiley.

Aniskowicz, B. T., H. Hamilton, D. R. Gray, and C. Downes. 1990. Nursing behavior of Arctic hares (*Lepus arcticus*). Pp. 643–64 in C. R. Harington, ed., *Canada's missing dimension: science and history in the Canadian Arctic islands.* Ottawa: Canadian Museum of Nature, 2.

Apfelbach, R., C. D. Blanchard, R. J. Blanchard, R. A. Hayes, and I. S. McGregor. 2005. The effects of predator odors in mammalian prey species: a review of field and laboratory studies. *Neurosci. and Biobehav. Rev.* 20:1123–44.

Asa, C. S., and L. D. Mech. 1995. A review of the sensory organs in wolves and their importance to life history. Pp. 287–91 in L. N. Carbyn, S. H. Fritts, and D. R. Seip, eds., *Ecology and conservation of wolves in a changing world.* Edmonton: Canadian Circumpolar Institute.

Atwell, G. 1964. Wolf predation on moose calf. *J. Mammal.* 45:313–14.

Atwood, T. C., E. M. Gese, and K. E. Kunkel. 2007. Comparative patterns of predation by cougars and recolonizing wolves in Montana's Madison Range. *J. Wildl. Mgmt.* 71:1098–1106.

———. 2009. Spatial partitioning of predation risk in a multiple predator–multiple prey system. *J. Wildl. Mgmt.* 73:876–84.

Audubon, J. J. (1843) 1897. *The Missouri River journals.* Vol. 1 in J. J. Audubon, M. R. Audubon, and E. Coues, eds., *Audubon and his journals.* New York: Charles Scribner's Sons.

Axelsson, E., A. Ratnakumar, M. Arendt, K. Maqbool, M. T. Webster, M. Perloski, O. Liberg, J. M. Arnemo, A. Hedhammar, and K. Lindblad-Toh. 2013. The genomic signature of dog domestication reveals adaptation to a starch-rich diet. *Nature* 495:360–64. doi: 10.1038/nature11837.

Bailey, T. N., and E. E. Bangs. 1980. Moose calving areas and use on the Kenai National Moose Range, Alaska. *Proc. of the North Am. Moose Conference Workshop* 16:289–313.

Ballard, W. 2011. Predator-prey relationships. Pp. 251–86 in D. G. Hewitt, ed., *Biology and management of white-tailed deer.* New York: CRC Press.

Ballard, W. B, T. H. Spraker, and K. P. Taylor. 1981. Causes of neonatal moose calf mortality in South Central Alaska. *J. Wildl. Mgmt.* 45:335–42.

Ballard, W. B., J. S. Whitman, and C. L. Gardner. 1987. Ecology of an exploited wolf population in south-central Alaska. *Wildl. Monogr.* no. 98. Bethesda, MD: Wildlife Society.

Banfield, A. W. F. 1954. *Preliminary investigation of the barren ground caribou.* Pt. 2. *Life history, ecology and utilization.* Wildlife Management Bulletin, ser. 1, no. 10B. Ottawa: Canada Dept. of Northern Affairs and National Resources, National Parks Branch, Canadian Wildlife Service.

Bangs, E. E., and S. H. Fritts. 1996. Reintroducing the gray wolf into central Idaho and Yellowstone National Park. *Wildl. Soc. Bull.* 24:402–13.

Barber-Meyer, S. M., and L. D. Mech. 2008. Factors influencing predation on juvenile ungulates and natural selection implications. *Wildl. Biol. in Practice* 4:8–29.

Barber-Meyer, S. M., L. D. Mech, and P. J. White. 2008. Elk calf survival and mortality following wolf restoration to Yellowstone National Park. *Wildl. Monogr.* no. 169.

Berger, J. 1979. Social ontogeny and behavioural diversity: consequences for bighorn sheep *Ovis canadensis* inhabiting desert and mountain environments. *J. Zool. Soc. Lond.* 192:251–66.

———. 2008. *The better to eat you with: fear in the animal world.* Chicago: University of Chicago Press.

Berger, J., J. E. Swenson, and I. Persson. 2001. Recolonizing carnivores and naïve prey: conservation lessons from Pleistocene extinctions. *Science* 291:1036–39.

Bergerud, A. T. 1980. A review of the population dynamics of caribou and wild reindeer in North America. Pp. 556–581 in E. Reimers, E. Gaare, and S. Skjenneberg, eds., *Proceedings of the Second International Reindeer/Caribou Symposium, Roros, Norway.* Trondeim, Norway: Direktoratetfo r vilto g ferskvannsfisk.

———. 1985. Antipredator strategies of caribou: dispersion along shorelines. *Can. J. Zool.* 63:1324–29.

Bergerud, A. T., H. E. Butler, and D. R. Miller. 1984. Antipredator tactics of calving caribou: dispersion in mountains. *Can. J. Zool.* 62:1566–75.

Bergerud, A. T., R. Ferguson, and H. E. Butler. 1990. Spring migration and dispersion of woodland caribou at calving. *Anim. Behav.* 39:360–68.

Bergerud, A. T., and R. E. Page. 1987. Displacement and dispersion of parturient caribou at calving as antipredator tactics. *Can. J. Zool.* 65:1597–1606.

Bertram, M. R., and M. T. Vivion. 2002. Moose mortality in eastern interior Alaska. *J. Wildl. Mgmt.* 66:747–58.

Best, T. L., and T. H. Henry. 1994. Lepus arcticus. *Mammal. Species* 457:1–9.

Bibikov, D. I. 1982. Wolf ecology and management in the USSR. Pp. 120–33 in F. H. Harrington, and P. C. Paquet, eds., *Wolves of the world: perspectives of behavior, ecology, and conservation.* Park Ridge, NJ: Noyes Publications.

Bjorge, R. R., and J. R. Gunson. 1983. Wolf predation of cattle on the Simonette River pastures in west-central Alberta. Pp. 106–11 in L. N. Carbyn, ed., *Wolves in Canada and Alaska: their status, biology, and management.* Report Series, no. 45. Edmonton: Canadian Wildlife Service.

Blood, D. A., D. W. Wishart, and D. R. Flook 1970. Weights and growth of Rocky Mountain bighorn sheep in western Alberta. *J. Wildl. Mgmt.* 34:451–55.

Boonstra, R. 2012. Reality as the leading cause of stress: rethinking the impact of chronic stress in nature. *Functional Ecol.* 27:11–23. doi: 10.1111/1365-2435.12008.

Boyd, D. K., R. R. Ream, D. H. Pletscher, and M. W. Fairchild. 1994. Prey taken by colonizing wolves and hunters in the Glacier National Park area. *J. Wildl. Mgmt.* 58:289–95.

Braitenberg, V. 1986. *Vehicles: experiments in synthetic psychology.* Cambridge, MA: MIT Press.

Bresler, D. E., G. Ellison, and S. Zamenhof. 1975. Learning deficits in rats with malnourished grandmothers. *Developmental Psych.* 8:315–23.

Brock, V. E., and R. H. Riffenburgh. 1960. Fish schooling: a possible factor in reducing predation. *J. du Conseil/Conseil Permanent International pour l'Exploration de la Mer* 25:307–17.

Bro-Jørgensen, J. 2013. Evolution of sprint speed in African savannah herbivores in relation to predation. *Evol.* 67 (11): 3371–76.

Brown, J. S., J. W. Laundre, and M. Gurung. 1999. The ecology of fear: optimal foraging, game theory, and trophic interactions. *J. Mammal.* 80:385–99.

Burkholder, B. L. 1959. Movements and behavior of a wolf pack in Alaska. *J. Wildl. Mgmt.* 23:1–11.

Buskirk, S. W., and P. S. Gipson. 1978. Characteristics of wolf attacks on moose in Mount McKinley National Park, Alaska. *Arctic* 31:499–502.

Butler, M. J., W. B. Ballard, and H. A. Whitlaw. 2006. Physical characteristics, hematology, and serum chemistry of free-ranging gray wolves, *Canis lupus*, southcentral Alaska. *Can. Field Nat.* 120:205–12.

Byers, J. A. 1997. *American pronghorn: social adaptations and the ghosts of predators past.* Chicago: University of Chicago Press.

Bygren, L. O., G. Kaati, and S. Edvinsson. 2001. Longevity determined by paternal ancestors' nutrition during their slow growth period. *Acta Biotheoretica* 49:53–59.

Cabot, W. B. 1920. Labrador. Boston: Small, Maynard and Co.

Caraco, T. 1981. Energy budgets, risk and foraging preferences in dark-eyed juncos (*Junco hymelais*). *Behav. Ecol. and Sociobiol.* 8:820–30.

Carbyn, L. N. 1974. *Wolf predation and behavioural interactions with elk and other ungulates in an area of high prey diversity.* Edmonton: Canadian Wildlife Service, Department of the Environment.

———. 1983. Wolf predation on elk in Riding Mountain National Park, Manitoba. *J. Wildl. Mgmt.* 47:963–76.

———. 2003. *The buffalo wolf: predators, prey and the politics of nature.* Washington, DC: Smithsonian Books.

Carbyn, L. N., S. M. Oosenbrug, and D. W. Anions. 1993. *Wolves, bison and the dynamics related to the Peace-Athabasca Delta in Canada's Wood Buffalo National Park.* Circumpolar Research Ser., no. 4. [Edmonton]: Canadian Circumpolar Institute.

Carbyn, L. N., and T. Trottier. 1987. Responses of bison on their calving grounds to predation by wolves in Wood Buffalo National Park. *Can. J. Zool.* 65:2072–78.

———. 1988. Descriptions of wolf attacks on bison calves in Wood Buffalo National Park, Canada. *Arctic* 41:297–302.

Carlson-Voiles, P. 2012. Glassy water morning. *International Wolf Magazine* 22 (2): 20–21.

Caro, T. M., L. Ombardo, A. W. Goldizen, and M. Kelly. 1995. Tail-flagging and other antipredator signals in white-tailed deer: new data and synthesis. *Behav. Ecol.* 6:442–50. doi: 10.1093/beheco/6.4.442.

Carstensen, M., G. D. DelGiudice, B. A. Sampson, and D. W. Kuehn. 2009. Survival, birth characteristics, and cause-specific mortality of white-tailed deer neonates. *J. Wildl. Mgmt.* 73:175–83.

Catlin, G. (1844) 2004. *North American Indians.* Reprint. New York: Penguin Books.

Cederlund, G., F. Sandegren, and K. Larsson. 1987. Summer movements of female moose and dispersal of their offspring. *J. Wildl. Mgmt.* 51:342–52.

Chandra, R. K. 1975. Antibody formation in first and second generation offspring of nutritionally deprived rats. *Science* 190:289–90.

Charnov, E. L., G. H. Orians, and K. Hyatt. 1976. Ecological implications of resource depression. *Am. Nat.* 110:247–59.

Child, K. N., K. K. Fujino, and M. W. Warren. 1978. A gray wolf (*Canis lupus columbianus*) and stone sheep (*Ovis dalli stonei*) fatal predator-prey encounter. *Can. Field Nat.* 92:399–401.

Childress, M. J., and M. A. Lung. 2003. Predation risk, gender and the group size effect: does elk vigilance depend upon the behaviour of conspecifics? *Anim. Behav.* 66:389–98. doi: 10.1006/anbe.2003.2217.

Christianson, D., and S. Creel. 2010. A nutritionally mediated risk effect of wolves on elk. *Ecol.* 91:1184–91.

Clark, K. R. F. 1971. Food habits and behavior of the tundra wolf on central Baffin Island. PhD thesis, University of Toronto.

Clarke, C. H. 1936. Moose seeks shelter for young. *Can. Field Nat.* 50:67–68.

Cook, R. A., J. G. Cook and L. D. Mech. 2004. Nutritional condition of northern Yellowstone elk. *J. Mammal.* 85:714–22.

Costain, B. 1989. Habitat use patterns and population trends among Shiras moose in a heavily logged region of northwest Montana. Master's thesis, University of Montana.

Cote, S. D., A. Peracino, and G. Simard. 1997. Wolf, *Canis lupus*, predation and maternal defensive behavior in Mountain Goats, *Oreamnos americanus. Can. Field Nat.* 111:389–92.

Cowan, I. M. 1947. The timber wolf in the Rocky Mountain National Parks in Canada. *Can. J. Res.* 25:139–74.

Creel, S., D. Christianson, S. Liley and J. A. Winnie Jr. 2007. Predation risk affects reproductive physiology and demography of elk. *Science* 315:960.

Creel, S., and J. A. Winnie Jr. 2005. Responses of elk herd size to fine-scale spatial and temporal variation in the risk of predation by wolves. *Anim. Behav.* 69:1181–89.

Creel, S., J. A. Winnie Jr., and D. Christianson. 2009. Glucocorticoid stress hormones and the effect of predation risk on elk reproduction. *PNAS* 106:12388–93.

Creel, S., J. A. Winnie Jr., D. Christianson, and S. Liley. 2008. Time and space in general model of antipredator response: tests with wolves and elk. *Anim. Behav.* 76:1139–46.

Creel, S., J. Winnie Jr., B. Maxwell, K. Hamlin, and M. Creel. 2005. Elk alter habitat selection as an antipredator response to wolves. *Ecol.* 86:3387–97.

Crisler, L. 1956. Observations of wolves hunting caribou. *J. Mammal.* 37:337–46.

Cubaynes, S., D. R. MacNulty, D. R., Stahler, K.A. Quimby, D. W. Smith, and T. Coulson. 2014. Density-dependent intraspecific aggression regulates survival in Yellowstone wolves. *J. Anim. Ecol.* 83:1344–56. doi: 10.1111/1365-2656.12238

Dai, A. 2011. Drought under global warming: a review. Wiley Interdisciplinary Reviews. *Climate Change* 2.1:45–65.

Darimont, C. T., and T. E. Reimchen. 2002. Intra-hair stable isotope analysis implies seasonal shift to salmon in gray wolf diet. *Can. J. Zool.* 80:1638–42.

Dauphine, T. C. 1969. A wolf kills a caribou calf. *Blue Jay* 27 (2): 99.

Dawkins, R., and Krebs, J. R. 1979. Arms races between and within species. *Proc. Royal Soc. Lond.*, ser. B 205:489–511.

DelGiudice, G. D. 1998. Surplus killing of white-tailed deer by wolves in northcentral Minnesota. *J. Mammal.* 79:227–35.

DelGiudice, G. D., J. Fieberg, M. R. Riggs, M. Carstensen Powell, and W. Pan. 2006. A long-term age-specific survival analysis of female white-tailed deer. *J. Wildl. Mgmt.* 70:1556–68.

DelGiudice, G. D., K. R. McCaffery, D. E. Beyer Jr., and M. E. Nelson. 2009. Prey of wolves in the Great Lakes region. Pp. 155–73 in T. R. Van Deelen and E. J. Heske, eds., *Recovery of gray wolves in the Great Lakes Region of the United States: an endangered species success story.* New York: Springer. doi: 10.1007/978-0-387-85952-1_10.

DelGiudice, G. D., L. D. Mech, K. E. Kunkel, E. M. Gese, and U. S. Seal. 1992. Seasonal patterns of weight, hematology, and serum characteristics of free-ranging female white-tailed deer in Minnesota. *Can. J. Zool.* 70:974–83.

DelGiudice, G. D., M. R. Riggs, P. Joly, and W. Pan. 2002. Winter severity, survival, and cause-specific mortality of female white-tailed deer in north-central Minnesota. *J. Wildl. Mgmt.* 66:698–717.

Demma, D. J., S. M. Barber-Meyer, and L. D. Mech. 2007. Testing use of Global Position System telemetry to study wolf predation on deer fawns. *J. Wildl. Mgmt.* 71:2767–75.

Demma, D. J., and L. D. Mech. 2009a. Wolf use of summer territory in northeastern Minnesota. *J. Wildl. Mgmt.* 72:380–84.

———. 2009b. Wolf, *Canis lupus*, visits to white-tailed deer, *Odocoileus virginianus*, summer ranges: optimal foraging? *Can. Field Nat.* 123:299–303.

Dill, L. M. 1987. Animal decision making and its ecological consequences: the future of aquatic ecology and behavior. *Can. J. Zool.* 65:803–11.

Dodge, R. I. 1878. *The hunting grounds of the great west, a description of the plains, game, and Indians of the great North American desert.* London: Chatto and Windus.

Dimond, S., and J. Lazarus. 1974. The problem of vigilance in animal life. *Brain, Behav. and Evol.* 9:60–79.

Fau, J. F., and I. R. Tempany. 1976. Wolf observations from aerial bison surveys, 1972–76. Pp. 11–14 in J. G. Stelfox, ed., *Wood Buffalo National Park: bison research annual report.* First Annual Report. Edmonton: Canadian Wildlife Services, Parks Canada.

Feldhamer, G. A., T. P. Kilbane, and D. W. Sharp. 1989. Cumulative effect of winter on acorn yield and deer body weight. *J. Wildl. Mgmt.* 53:292–95.

Forester, J. D., A. R. Ives, M. G. Turner, D. P. Anderson, D. Fortin, H. L. Beyer, D. W. Smith, and M. S. Boyce. 2007. State-state models link elk movement patterns to landscape characteristics in Yellowstone National Park. *Ecol. Appl.* 77:285–99.

Fox, J. L., and G. P. Streveler. 1986. Wolf predation on mountain goats in southeastern Alaska. *J. Mammal.* 67:192–95.

Fox, M. W. 1969. Ontogeny of prey-killing in *Canidae. Behav.* 35:259–72.

Frame, P. F., D. S. Hik, H. D. Cluff, and P. C. Paquet. 2004. Long foraging movement of a denning tundra wolf. *Arctic* 57:196–203.

Frank, H., M. G. Frank, and L. M. Hasselbach. 1989. Motivation and insight in wolf (*Canis lupus*) and Alaskan malamute (*Canis familiaris*): visual discrimination learning. *Bull. Psychon. Soc.* 27:455–58.

Franzmann, A. W., C. C. Schwartz, and R. O. Peterson. 1980. Moose calf mortality in summer on the Kenai Peninsula, Alaska. *J. Wildl. Mgmt.* 44:764–68.

Fremont, J. C. 1845. *Report of the exploring expedition to the Rocky Mountains in the year 1842, and to Oregon and North Carolina in the years 1843–44.* Washington, DC: Gales and Seaton.

Frijlink, J. H. 1977. Patterns of wolf pack movements prior to kills as read from tracks in Algonquin Provincial Park, Ontario, Canada. *Bijdrafen tot de dierkunde* 47:131–37.

Fritts, S. H., and L. D. Mech. 1981. Dynamics, movements, and feeding ecology of a newly protected wolf population in northwestern Minnesota. *Wildl. Monogr.* 80:1–79.

Fryxell, J. M., A. Mosser, A. R. E. Sinclair, and C. Packer. 2007. Group formation stabilizes predator-prey dynamics. *Nature* 449:1041–44.

Fuller, T. K., and L. B. Keith. 1980. Wolf population dynamics and prey relationships in northeastern Alberta. *J. Wildl. Mgmt.* 44:583–602.

Fuller, T. K., L. D. Mech, and J. Fitts-Cochran. 2003. Population dynamics. Pp. 161–91 in L. D. Mech and L. Boitani, eds., *Wolves: behavior, ecology, and conservation.* Chicago: University of Chicago Press.

Fuller, W. A. 1960. Behaviour and social organization of the wild bison of Wood Buffalo National Park, Canada. *Arctic* 13 (1): 3–19.

———. 1962. *The biology and management of bison of Wood Buffalo National Park.* Wildlife Management Bulletin, ser. 1, no. 16. Ottawa: Canadian Wildlife Service.

Galton, F. 1871. Gregariousness in cattle and men. *MacMillan's Magazine* 23:353.

Garland, T., Jr. 1983. The relation between maximal running speed and body mass in terrestrial mammals. *J. Zool. Soc. Lond.* 199:157–70.

Garrott, R. A., P. J. White, and J. J. Rotella. 2009. The Madison headwaters elk herd: transitioning from bottom-up regulation to top-down limitation. Pp. 489–517 in R. A. Garrott, P. J. White, and F. G. R. Watson, eds., *The ecology of large mammals in central Yellowstone: sixteen years of integrated field studies.* San Diego, CA: Elsevier.

Garroway, C. J., and H. G. Broders. 2005. The quantitative effects of population density and winter weather on the body condition of white-tailed deer (*Odocoileus virginianus*) in Nova Scotia, Canada. *Can. J. Zool.* 83:1246–1256. doi: 10.1139/Z05–118.

Gasaway, W. C., R. D. Boertje, D. V. Grangaard, D. G. Kelleyhouse, R. O. Stephenson, and D. G. Larsen. 1992. The role of predation in limiting moose at low densities in Alaska and Yukon and implications for conservation. *Wildl. Monogr.* no. 120.

Geist, V. 1971. *Mountain sheep: a study in behavior and evolution.* Chicago: University of Chicago Press.

———. 1999. Adaptive strategies in American mountain sheep: effects of climate, latitude and altitude, ice age evolution, and neonatal security. Pp. 192–208 in R. Valdez and P. R. Krausman, eds., *Mountain sheep of North America.* Tucson: University of Arizona Press.

Geremia, C., P. J. White, R. L. Wallen, F. G. R. Watson, J. J. Treanor, J. J. Borkowski, C. S. Potter, and R. L. Crabtree. 2011. Predicting bison migration out of Yellowstone National Park using Bayesian Models. *PLoS ONE* 6:1–9.

Gower, C. N., R. A. Garrott, and P. J. White. 2009b. Elk foraging behavior: does predation risk reduce time for food acquisition? Pp. 423–50 in R. A. Garrott, P. J. White, and F. G. R. Watson, eds., *The ecology of large mammals in central Yellowstone: sixteen years of integrated field studies.* New York: Elsevier Academic Press.

Gower, C.N. R. A. Garrott, P. J. White, S. Cherry, and N. G. Yoccoz. 2009a. Elk group size and wolf predation: a flexible strategy when faced with variable risk. Pp. 401–22 in R. A. Garrott, P. J. White, and F. G. R. Watson, eds., *The ecology of large mammals in central Yellowstone: sixteen years of integrated field studies.* New York: Elsevier Academic Press.

Gray, D. R. 1970. The killing of a bull muskox by a single wolf. *Arctic* 23:197–99.

———. 1983. Interactions between wolves and muskoxen on Bathurst Island, NWT, Canada. *Acta Zool. Fenn.* 174:255–57.

———. 1987. *The muskoxen of Polar Bear Pass.* Markham, Ontario: National Museum of Natural Sciences, Fitzhenry and Whiteside.

Gude, J. A., R. A. Garrott, J. J. Borkowski, and F. King. 2006. Prey risk allocation in a grazing ecosystem. *Ecol. Appl.* 16:285–98.

Gula, R. 2004. Influence of snow cover on wolf predation patterns in Bieszczady Mountains, Poland. *Wildl. Biol.* 10:17–23.

Guthrie, R. D. 2006. New carbon dates link climatic change with human colonization and Pleistocene extinctions. *Nature* 441:207–9.

Haber, G. C. 1968. The social structure and behavior of an Alaskan wolf population. Master's thesis. Northern Michigan University, Marquette.

———. 1977. Socio-ecological dynamics of wolves and prey in a subarctic ecosystem. PhD thesis. University of British Columbia.

Hagar, J. A., Marshfield Hills, MA. 1968. Personal correspondence to L. D. Mech.

Hall, A. M. 1971. Ecology of beaver and selection of prey by wolves in central Ontario. Master's thesis. University of Toronto.

Hamilton, W. D. 1971. Geometry for a selfish herd. *J. Theoretical Biol.* 31:295–311.

Hamlin, K. L., R. A. Garrott, P. J. White and J. A. Cunningham. 2009. Contrasting wolf-ungulate interactions in the greater Yellowstone ecosystem. Pp. 541–78 in R. A. Garrott, P. J. White, and F. G. R. Watson, eds., *The ecology of large mammals in central Yellowstone: sixteen years of integrated field studies.* New York: Elsevier Academic Press.

Haugen, H. S. 1987. Den-site behavior, summer diet, and skull injuries of wolves in Alaska. Master's thesis. University of Alaska, Fairbanks.

Hayes, R. D. 1995. Numerical and functional responses of wolves and regulation of moose in the Yukon. Master's thesis. Simon Fraser University, Burnaby, BC.

Hayes, R. D., A. M. Baer, and D. G. Larsen. 1991. *Population dynamics and prey relationships of an exploited and recovering wolf population in the southern Yukon.* Yukon Fish & Wildlife Branch, Dept. of Renewable Resources, Final Report TR-91-1.

Heard, D. C. 1992. The effect of wolf predation and snow cover on musk-ox group size. *Am. Nat.* 139:190–204.

Hebblewhite, M. 2000. Wolf and elk predator-prey dynamics in Banff National Park. Master's thesis. School of Forestry, University of Montana.

Hebblewhite, M., E. H. Merrill, and T. L. McDonald. 2005. Spatial decomposition of predation risk using resource selection functions: an example in a wolf-elk predator-prey system. *Oikos* 111:101–11.

Hebblewhite, M., and D. H. Pletscher. 2002. Effects of elk group size on predation by wolves. *Can. J. Zool.* 80:800–809.

Heffelfinger, J. R. 2011. Taxonomy, evolutionary history, and distribution. Pp. 3–39 in D. G. Hewitt, ed., *Biology and management of white-tailed deer.* New York: CRC Press.

Heimer, W. 1973. Dall sheep movement and mineral lick use. App. 4. Alaska Department of Fish and Game, Project Report Number W-17-2-5.

Hoefs, M., H. Hoefs, and D. Burles. 1986. Observations on Dall sheep, *Ovis dalli dalli*–grey wolf, *Canis lupus pambasileus*, interactions in the Kluane Lake area, Yukon. *Can. Field Nat.* 100:78–84.

Holling, C. S. 1965. The functional response of predators to prey density and its role in mimicry and population regulation. *Mem. Entomol. Soc. of Can.* 45:1–62.

Hone, E. 1934. The present status of the muskox in Arctic North America and Greenland. *Special Publication, American Committee for International Wild Life Protection*, no. 5. Philadelphia: Wilderness Club.

Hoogland, J. L. 1979. The effect of colony size on individual alertness of prairie dogs (Schiuridae: *Cynomys* spp.). *Anim. Behav.* 27:394–407.

Horejsi, B. L. 1976. Suckling and feeding behaviour in relation to lamb survival in bighorn sheep (*Ovis Canadensis*). PhD diss. University of Calgary, Alberta.

Hoskinson, R. L., and L. D. Mech. 1976. White-tailed deer migration and its role in wolf predation. *J. Wildl. Mgmt.* 40:429–41.

Howe, P. E., H. A. Mattill, and P. B. Hawk. 1912. Fasting studies. 4. Distribution of nitrogen during a fast of one hundred and seventeen days. *J. Biol. Chem.* 11:103–27.

Hubert, B. A. 1974. Estimated productivity of muskoxen (*Ovibos moschatus*) on northeastern Devon Island, NWT. Master's thesis. University of Manitoba, Winnipeg.

Hudson, R. J., J. C. Haigh, and A. B. Bubenik. 2002. Physical and physiological adaptations. Pp. 199–258 in D. E. Toweill and J. W. Thomas, eds., *North American elk: ecology and management*. Washington, DC: Smithsonian Institution Press.

Huggard, D. J. 1993a. Prey selectivity of wolves in Banff National Park. Pt. 1. Prey species. *Can. J. Zool.* 71:130–39.

———. 1993b. Prey selectivity in Banff National Park. Pt. 2. Age, sex, and condition of elk. *Can. J. Zool.* 71:140–41.

Hummel, M., and J. C. Ray. 2008. Caribou and the North: a shared future. Toronto: Dundurn Press.

Ivlev, V. S. 1961. *Experimental ecology of the feeding of fishes.* New Haven, CT: Yale University Press.

Jackson, W. H. 1942. Three unlucky wolves. *Alaska Sportsman* 8 (7): 8–9, 29–30.

James, A. R. C., and A. K. Stuart-Smith. 2000. Distribution of caribou and wolves in relation to linear corridors. *J. Wildl. Mgmt.* 64 (1): 154–59.

Jedrzejewski, W., K. Schmidt, J. Theuerkauf, B. Jedrzejewski, and H. Okarma. 2001. Daily movements and territory use by radio-collared wolves (*Canis lupus*) in Bialowieza Primeval Forest in Poland. *Can. J. Zool.* 79:1993–2004.

Jordan, P. A., R. O. Peterson, and K. A. LeDoux. 2010. Swimming wolves, *Canis lupus,* attack a swimming moose, *Alces alces. Can. Field Nat.* 124:54–56.

Joslin, P. W. B. 1966. Summer activities of two timber wolf (*Canis lupus*) packs in Algonquin Park. Master's thesis. University of Toronto.

Kaati, G., L. O. Bygren, M. Pembrey, and M. Sjostrom. 2007. Transgenerational response to nutrition, early life circumstances and longevity. *Eur. J. Hum. Genetics* 15:784–90.

Kauffman, M. J., N. Varley, D. W. Smith, D. R. Stahler, D. R. MacNulty and M. S. Boyce. 2007. Landscape heterogeneity shapes predation in a newly restored predator-prey system. *Ecol. Lett.* 10:690–700.

Kelsall, J. P. 1957. *Continued barren-ground caribou studies.* Wildlife Management Bulletin, ser. 1, no. 12. Ottawa: Dept. of Northern Affairs and National Resources, National Parks Branch, Canadian Wildlife Service.

———. 1960. *Co-operative studies of barren-ground caribou, 1957–58.* Wildlife Management Bulletin, ser. 1, no. 15. Ottawa: Dept. of Northern Affairs and National Resources, National Parks Branch, Canadian Wildlife Service.

———. 1968. *The caribou: the migratory barren-ground caribou of Canada.* Canadian Wildlife Service Monograph Series, no. 3. Ottawa: Queen's Printer.

Kingsolver, J. G., H. E. Hoekstra, J. M. Hoekstra, D. Berrigan, S. N. Vignieri, C. E. Hill, A. Hoang, P. Gibert, and P. Beerli. 2001. The strength of phenotypic selection in natural populations. *Am. Nat.* 157:245–61.

Kiss, B. W., S. K. Johnstone, and R. P. Berger. 2010. Predation of a barren-ground caribou, *Rangifer tarandus groenlandicus*, by a single gray wolf, *Canis lupus*, in northern Manitoba. *Can. Field. Nat.* 124:270–71.

Kittle, A. M., J. M. Fryxell, G. E. Desy, and J. Hamr. 2008. The scale-dependent impact of wolf predation risk on resource selection by three sympatric ungulates. *Oecologia* 157:163–75. doi: 10.1007/s00442-008-1051-9.

Kolenosky, G. B. 1972. Wolf predation on wintering deer in east-central Ontario. *J. Wildl. Mgmt.* 36:357–69.

Krebs, J. R. 1980. Optimal foraging, predation risk, and territory defense. *Ardea* 68:83–90.

Kruuk, H. 1972. Surplus killing by carnivores. *J. Zool.* 166:233–44.

Kunkel, K. E., and L. D. Mech. 1994. Wolf and bear predation on white-tailed deer fawns. *Can. J. Zool.* 72:1557–65.

Kunkel, K., and D. H. Pletscher. 1999. Species-specific population dynamics of cervids in a multipredator ecosystem. *J. Wildl. Mgmt.* 63:1082–93.

———. 2000. Habitat affecting vulnerability of moose to predation by wolves in southeastern British Columbia. *Can. J. Zool.* 72:1557–65.

———. 2001. Winter hunting patterns of wolves in and near Glacier National Park, Montana. *J. Wildl. Mgmt.* 65:520–30.

Kunkel, K. E., D. H. Pletscher, D. K. Boyd, R. R. Ream, and M. W. Fairchild. 2004. Factors correlated with foraging behavior of wolves in and near Glacier National Park, Montana. *J. Wildl. Mgmt.* 68:167–78.

Kunkel, K. E., T. K. Ruth, D. H. Pletscher, and M. G. Hornocker. 1999. Winter prey selection by wolves and cougars in and near Glacier National Park, Montana. *J. Wildl. Mgmt.* 63:901–10.

LaGory, K. E. 1986. Habitat, group size, and the behaviour of white-tailed deer. *Behav.* 98:168–79.

———. 1987. The influence of habitat and group characteristics on the alarm and flight response of white-tailed deer. *Anim. Behav.* 35:20–25.

Landis, R. 1998. *Yellowstone wolves—predation.* Gardiner, MT: Trailwood-Landis Films.

Langley, M. A., and D. H. Pletscher. 1994. Calving areas of moose in northwestern Montana and southwestern British Columbia. *Alces* 30:127–35.

Larsen, D. G., D. A. Gauthier, and R. L. Markel. 1989. Causes and rate of moose mortality in the southwest Yukon. *J. Wildl. Mgmt.* 53:548–57.

Latham, A. D. M., M. C. Latham, N. A. Mccutchen, and S. Boutin. 2011. Invading white-tailed deer change wolf-caribou dynamics in northeastern Alberta. *J. Wildl. Mgmt.* 75:204–12. doi: 10.1002/jwmg.28

Latour, P. 1987. Observations on demography, reproduction and morphology of muskoxen (*Ovibos moschatus*) on Banks Island, Northwest Territories. *Can. J. Zool.* 65:265–69. 10.1139/z87-041.

Laundre, J. W., L. Hernandez, and K. B. Altendorf. 2001. Wolves, elk, and bison: reestablishing the "landscape of fear" in Yellowstone National Park, U.S.A. *Can. J. Zool.* 79:1401–9.

Lent, P. C. 1978. Musk-ox. Pp. 135–47 in J. L. Schmidt and D. L. Gilbert, eds., *Big game of North America.* Harrisburg, PA: Wildlife Management Institute, Stackpole Books.

Leptich, D. J., and J. R. Gilbert. 1986. Characteristics of moose calving sites in northern Maine as determined by multivariate analysis: a preliminary investigation. *Alces* 22:69–81.

Lerman, K., and A. Galstyan. 2002. Mathematical model of foraging in a group of robots: effect of interference. *Autonomous Robots* 13:127–41.

Lewis, M. A., and J. D. Murray. 1993. Modelling territoriality and wolf-deer interactions. *Nature* 366:738–40.

Liley, S., and S. Creel. 2008. What best explains vigilance in elk: characteristics of prey, predators, or the environment? *Behav. Ecol.* 19:245–54.

Lima, S. L., and L. M. Dill. 1990. Behavioral decisions made under the risk of predation: a review and prospectus. *Can. J. Zool.* 68:619–40

Lingle, S., and S. M. Pellis. 2002. Fight or flight? antipredator behavior and the escalation of coyote encounters with deer. *Oecologia* 131:154–64.

Link, R. 2004. "Elk" (http://wdfw.wa.gov/living/elk.pdf). Adapted from his *Living with Wildlife in the Pacific Northwest.* Seattle: University of Washington Press.

Llaneza, L., A. Fernandez, and C. Nores. 1996. Dieta del lobo en dos zonas de Asturias que difieren en carga ganadera [Wolf diet in two areas of Asturias with different livestock abundance]. *Doñana, Acta Vertebrata* 23:201–14.

Lopez-Bao, J. V., V. Sazatornil, L. Llaneza, and A. Rodriguez. 2013. Indirect effects on heathland conservation and wolf persistence of contradictory policies that threaten traditional free-ranging horse husbandry. *Conserv. Lett.* 00:1–8. doi: 10.1111/con1.12014

Lott, D. F. 2002. *American bison: a natural history.* Berkeley: University of California Press.

Lumey, L. H., and A. D. Stein. 1997. Offspring birth weights after maternal intrauterine undernutrition: a comparison within sibships. *Am. J. Epidem.* 146:810–19.

MacArthur, R. A., R. H. Johnston, and V. Geist. 1979. Factors influencing heart rate in free-ranging bighorn sheep: a physiological approach to the study of wildlife harassment. *Can. J. Zool.* 57:2010–21.

MacArthur, R. H., and E. R. Pianka. 1966. On optimal use of a patchy environment. *Am. Nat.* 100:603–9.

MacFarlane, R. 1905. Notes on mammals collected and observed in the northern Mackenzie River district, Northwest Territories of Canada, with remarks on explorers and explorations of the far north. *Proc. U.S. Natl. Mus.* 28:673–764.

MacNulty, D. R. 2002. The predatory sequence and the influence of injury risk on hunting behavior in the wolf. MS thesis. University of Minnesota, St. Paul.

———. 2007. Development, senescence, and cooperation in the predatory behavior of a social carnivore, *Canis lupus.* PhD diss. University of Minnesota, St. Paul.

MacNulty, D. R., L. D. Mech, and D. W. Smith. 2007. A proposed ethogram of large-carnivore predatory behavior, exemplified by the wolf. *J. Mammal.* 88:595–605.

MacNulty, D. R., and D. W. Smith. 2006. Bison foraging response to the risk of wolf predation in a spatially heterogeneous winter environment: a preliminary assessment. Pp. 3–10 in H. J. Harlow and M. Harlow, eds., *Annual report: University of Wyoming, National Park Service Research Center,* vol. 28. Laramie: The Center.

MacNulty, D. R., D. W. Smith, L. D. Mech, and L. E. Eberly. 2009b. Body size and predatory performance in wolves: is bigger better? *J. Anim. Ecol.* 78:532–39.

MacNulty, D. R., D. W. Smith, L. D. Mech, J. A. Vucetich, and C. Packer. 2012. Nonlinear effects of group size on the success of wolves hunting elk. *Behav. Ecol.* 23:75–82.

MacNulty, D. R., D. W. Smith, J. A. Vucetich, L. D. Mech, D. R. Stahler, and C. Packer. 2009a. Predatory senescence in ageing wolves. *Ecol. Lett.* 12:1–10.

MacNulty, D. R., A. Tallian, D. R. Stahler, and D. W. Smith. 2014. Influence of group size on the success of wolves hunting bison. *PLoS ONE* 9 (11): 112884. doi.10.137/journal.pone.0112884.

MacNulty, D. R., N. Varley, and D. W. Smith. 2001. Grizzly bear, *Ursus arctos,* usurps bison, *Bison bison,* captured by wolves, *Canis lupus,* in Yellowstone National Park, Wyoming. *Can. Field Nat.* 115:495–98.

Madden, J. D., R. C. Arkin, and D. R. MacNulty. 2010. Multi-robot system based on model of wolf hunting behavior to emulate wolf and elk interactions. Pp. 1043–50 in *IEEE-ROBIO 2010: 2010 IEEE International Conference on Robotics and Biomimetics: December 14–18, 2010, Tianjin, China.* Piscataway, NJ: IEEE.

Makridin, V. P. 1962. The wolf in the Yamal north. *Zool. Zh.* 41:1413–17. (Translation by Peter Lent.)

Mao, J. S., M. S. Boyce, D. W. Smith, F. J. Singer, D. J. Vales, J. M. Vore, and E. H. Merrill. 2005. Habitat selection by elk before and after wolf reintroduction in Yellowstone National Park. *J. Wildl. Mgmt.* 69:1691–1707.

Mataric, M. J. 1993. Designing emergent behaviors: from local interactions to collective intelligence. Pp. 432–41 in J.-A. Meyer, H. Roitblat, and S. Wilson, eds., *From animals to animats 2: Proceedings of the International Conference on Simulation of Adaptive Behavior.* Cambridge, MA: MIT Press.

McCullough, D. R. 1969. *The tule elk: its history, behavior, and ecology.* University of California Publications in Zoology, vol. 88. Berkeley: University of California Press.

McRoberts, R. E., L. D. Mech, and R. O. Peterson. 1995. The cumulative effect of consecutive winters' snow depth on moose and deer populations: a defense. *J. Anim. Ecol.* 64:131–35.

Mead, J. R. 1899. Some natural history notes of 1859. *Trans. Kans. Acad. Sci.* 16:280–81.

Meagher, M. 1973. *The bison of Yellowstone National Park.* Scientific Monograph Series, no. 1. Washington, DC: Department of the Interior, National Park Service.

———. 1986. *Bison bison. Mammal. Species,* no. 266.

Mech, L. D. 1966a. *The wolves of Isle Royale.* National Parks Fauna Series, no. 7. Washington, DC: Government Printing Office.

———. 1966b. Hunting behavior of timber wolves in Minnesota. *J. Mammal.* 47:347–48.

———. 1970. *The wolf: the ecology and behavior of an endangered species.* Garden City, NY: Natural History Press.

———. 1977a. Wolf pack buffer zones as prey reservoirs. *Science* 198:320–21.

———. 1977b. Population trend and winter deer consumption in a Minnesota wolf pack. Pp. 55–83 in R. L. Phillips and C. Jonkel, eds., *Proceedings of the 1975 predator symposium.* Missoula: Montana Forest and Conservation Experimental Station.

———. 1977c. Productivity, mortality and population trend in wolves from northeastern Minnesota. *J. Mammal.* 58:559–74.

———. 1977d. Where can the wolf survive? *Natl. Geogr. Mag.* 152 (4): 518–37.

———. 1984. Predators and predation. Pp. 189–200 in K. L. Hall, ed., *White-tailed deer: ecology and management.* Harrisburg, PA: Stackpole Books.

———. 1987. At home with the Arctic wolf. *Natl. Geogr. Mag.* 171:562–93.

———. 1988. *The arctic wolf: living with the pack.* Stillwater, MN: Voyageur Press.

———. 1992. Winter daytime activity of wolves in northeastern Minnesota. *J. Mammal.* 73:570–71.

———. 1994a. Buffer zones of territories of gray wolves as regions of intraspecific strife. *J. Mammal.* 75:199–202.

———. 1994b. Regular and homeward travel speeds of arctic wolves. *J. Mammal.* 75:741–42.

———. 1995. A ten-year history of the demography and productivity of an arctic wolf pack. *Arctic* 48:329–32.

———. 1997. An example of endurance in an old wolf, *Canis lupus. Can. Field Nat.* 111:654–55.

———. 1999. Alpha status, dominance, and division of labor in wolf packs. *Can. J. Zool.* 77:1196–1203.

———. 2000. A record-large wolf pack in Minnesota. *Can. Field Nat.* 114:504–5.

———. 2006. Age-related body mass and reproductive measurements of gray wolves in Minnesota. *J. Mammal.* 87:80–84

———. 2007a. Femur-marrow fat of white-tailed deer fawns killed by wolves. *J. Wildl. Mgmt.* 71:920–23.

———. 2007b. Possible use of foresight, understanding, and planning by wolves hunting muskoxen. *Arctic* 60:145–49.

———. 2007c. Annual arctic wolf pack size related to arctic hare numbers. *Arctic* 60:309–11.

———. 2009. Long-term research on wolves in the Superior National Forest. Pp. 15–34 in A. P. Wydeven, E. J. Heske, and T. R, Van Deelen, eds., *Recovery of gray wolves in the Great Lakes Region of the United States: an endangered species success story.* New York: Springer.

———. 2010. Proportion of calves and adult muskoxen killed by wolves in summer on Ellesmere Island. *Can. Field Nat.* 124:258–60.

———. 2011. Gray wolf (*Canis lupus*) movements and behavior around a kill site and implications for GPS collar studies. *Can. Field Nat.* 125 (4): 353–56.

———. 2013. The challenge of wolf recovery: an ongoing dilemma for state managers. *Wildl. Professional* 7 (1): 32–37.

———. 2014. A gray wolf (*Canis lupus*) delivers live prey to a pup. *Can. Field Nat.* 128 (1): 189–90.

Mech, L. D., and L. G. Adams. 1999. Killing of a muskox, *Ovibos moschatus,* by two wolves, *Canis lupus,* and subsequent caching. *Can. Field Nat.* 113:673–75.

Mech, L. D., L. G. Adams, T. J. Meier, J. W. Burch, and B. W. Dale. 1998. *The wolves of Denali.* Minneapolis: University of Minnesota Press.

Mech, L. D., and L. Boitani, eds. 2003. *Wolves, behavior, ecology, and conservation.* Chicago: University of Chicago Press.

Mech, L. D., and H. D. Cluff. 2011. Movements of wolves at the northern extreme of the species' range including during four months of darkness. *PLoS ONE* 6 (10): e25328.

Mech, L. D., and G. D. DelGiudice. 1985. Limitations of the marrow-fat technique as an indicator of condition. *Wildl. Soc. Bull.* 13:204–6.

Mech, L. D., and L. D. Frenzel Jr. 1971. *Ecological studies of the timber wolf in northeastern Minnesota.* USDA Forest Service Research Paper NC-52. St. Paul: US North Central Forest Experimental Station.

Mech, L. D., L. D. Frenzel Jr., and P. D. Karns. 1971. The effect of snow conditions on the ability of wolves to capture deer. Pp. 51–59 in L. D. Mech and L. D. Frenzel, Jr., eds., *Ecological studies of the timber wolf in northeastern Minnesota.* USDA Forest Service Research Paper NC-52. St. Paul: US North Central Forest Experimental Station.

Mech, L. D., and E. K. Harper. 2002. Differential use of a wolf, *Canis lupus,* pack territory edge and core. *Can. Field Nat.* 116:315–16.

Mech, L. D., and P. D. Karns. 1977. *Role of the wolf in a deer decline in the Superior National Forest.* USDA Forest Service Research Report NC-148. St. Paul: US North Central Forest Experimental Station.

Mech, L. D., and M. Korb. 1978. An unusually long pursuit of a deer by a wolf. *J. Mammal.* 59:860–61.

Mech, L. D., and R. E. McRoberts. 1990. Survival of white-tailed deer fawns in relation to maternal age. *J. Mammal.* 71:465–67.

Mech, L. D., R. E. McRoberts, R. O. Peterson, and R. E. Page. 1987. Relationship of deer and moose populations to previous winters' snow. *J. Anim. Ecol.* 56:615–28.

Mech, L. D., and M. E. Nelson. 1990. Evidence of prey-caused mortality in three wolves. *Am. Midl. Nat.* 123:207–8.

———. 2013. Age structure of Moose (*Alces alces*) killed by Gray Wolves (*Canis lupus*) in northeastern Minnesota, 1967–2011. *Can. Field Nat.* 127:70–71

Mech, L. D., M. E. Nelson, and R. E. McRoberts. 1991. Effects of maternal and grandmaternal nutrition on deer mass and vulnerability to wolf predation. *J. Mammal.* 72:146–51.

Mech, L. D., and R. O. Peterson. 2003. Wolf-prey relations.

Pp.131–57 in L. D. Mech and L. Boitani, eds., *Wolves: behavior, ecology, and conservation.* Chicago: University of Chicago Press.

Mech, L. D., U. S. Seal, and S. M. Arthur. 1984. Recuperation of a severely debilitated wolf. *J. Wildl. Dis.* 20 (2): 166–68.

Mech, L. D., D. W. Smith, K. M. Murphy, and D. R. MacNulty. 2001. Winter severity and wolf predation on a formerly wolf-free elk herd. *J. Wildl. Mgmt.* 65:998–1003.

Mech, L. D., and S. Tracy. 2004. Record high wolf, *Canis lupus*, pack density. *Can. Field Nat.* 118:127–29.

Messier, F. 1991. The significance of limiting and regulating factors on the demography of moose and white-tailed deer. *J. Anim. Ecol.* 60:377–93.

———. 1995. Is there evidence for a cumulative effect of snow on moose and deer populations? *J. Anim. Ecol.* 64:136–40.

Metz, M. C., D. W. Smith, J. A. Vucetich, D. R. Stahler, and R. O. Peterson. 2012. Seasonal patterns of predation for gray wolves in the multi-prey system of Yellowstone National Park. *J. Anim. Ecol.* 81:553–63.

Middleton, A. D., M. J. Kauffman, D. E. McWhirter, M. D. Jimenez, R. C. Cook, J. G. Cook, S. E. Albeke, H. Sawyer, and P. J. White. 2013. Linking anti-predator behavior to prey demography reveals limited risk effects of an actively hunting carnivore. *Ecol. Lett.* 16 (8): 1023–30. doi: 10.111/ele.12133.

Miller, F. L. 1975. Observations of wolf predation on barren ground caribou in winter. Pp. 209–20 in J. R. Luick, P. C. Lent, D. R. Klein, and R. G. White, eds., *Proceedings First International Reindeer and Caribou Symposium.* Biological Papers of the University of Alaska, Special Report no. 1. Fairbanks: University of Alaska.

Miller, F. L., A. Gunn, and E. Broughton. 1985. Surplus killing as exemplified by wolf predation on newborn caribou. *Can. J. Zool.* 63:295–300.

Molvar, E. M., and R. T. Bowyer. 1994. Costs and benefits of group living in a recently social ungulate: the Alaskan moose. *J. Mammal.* 75:621–30.

Monteith, K. L., L. E. Schmitz, J. A. Jenks, J. A. Delger, and R. T. Bowyer. 2009. Growth of male white-tailed deer: consequences of maternal effects. *J. Mammal.* 90:651–60.

Montgomery, R. A., J. A. Vucetich, R. O. Peterson, G. J. Roloff, and K. F. Millenbah. 2013. The influence of winter severity, predation and senescence on moose habitat use. *J. Anim. Ecol.* 82:301–309.

Mowat, F. 1963. *Never cry wolf.* New York: Dell.

Mukherjee, S., and M. R. Heithaus. 2013. Dangerous prey and daring predators: a review. *Biol. Rev.* 88:550–63. doi: 10.1111/brv.12014.

Muller-Schwarze, D. 1972. Responses of young black-tailed deer to predator odors. *J. Mammal.* 53:393–94.

Murie, A. 1944. *The wolves of Mount McKinley.* Fauna of the National Parks of the United States, Fauna Series, no. 5. Washington, DC: Government Printing Office.

Muro, C., R. Escobedo, L. Spector, and R. P. Coppinger. 2011. Wolf-pack (*Canis lupus*) hunting strategies emerge from simple rules in computational simulations. *Behav. Process.* 88:192–97.

Nelson, M. E., and L. D. Mech. 1981. Deer social organization and wolf depredation in northeastern Minnesota. *Wildl. Monogr.* 77:1–53.

———. 1984. Observation of a swimming wolf killing a swimming deer. *J. Mammal.* 6:143–44.

———. 1985. Observations of a wolf killed by a deer. *J. Mammal.* 66:187–88.

———. 1986a. *Deer population in the central Superior National Forest, 1967–1985.* USDA Forest Service Research Paper NC-271. St. Paul: US North Central Forest Experiment Station.

———. 1986b. Mortality of white-tailed deer in northeastern Minnesota. *J. Wildl. Mgmt.* 50:691–98.

———. 1987. Demes within a northeastern Minnesota deer population. Pp. 27–40 in B. D. Chepko-Sade and Z. Halpin, eds., *Mammalian dispersal patterns.* Chicago: University of Chicago Press.

———. 1990. Weights, productivity, and mortality of old white-tailed deer. *J. Mammal.* 71:689–91.

———. 1991. White-tailed deer movements and wolf predation risk. *Can. J. Zool.* 69:2696–99.

———. 1993. Prey escaping wolves despite close proximity. *Can. Field Nat.* 107:245–46.

———. 1994. A single deer stands off three wolves. *Am. Midl. Nat.* 131:207–8.

———. 1999. Twenty-year home range dynamics of a white-tailed deer matriline. *Can. J. Zool.* 77:1128–35.

———. 2000. Proximity of white-tailed deer, *Odocoileus virginianus*, home ranges to wolf, *Canis lupus*, pack homesites. *Can. Field Nat.* 114:503–4.

———. 2006. Causes of a 3-decade dearth of deer in a wolf-dominated ecosystem. *Am. Midl. Nat.* 155:373–82.

Newman, J. A., and T. Caraco. 1987. Foraging, predation hazard and patch use in gray squirrels. *Anim. Behav.* 35:1804–13.

Nichols, L. 1978. Dall's sheep. Pp. 173–89 in J. L. Schmidt and D. L. Gilbert, eds., *Big game of North America: ecology and management.* Harrisburg, PA: Stackpole Books.

Nowak, R. M. 1991. *Walker's mammals of the world.* Vol. 2. 5th ed. Baltimore, MD: Johns Hopkins University Press.

Olson, S. F. 1938. Organization and range of the pack. *Ecology* 19:168170.

Osborne, T. O., T. F. Paragi, J. L. Bodkin, A. J. Loranger, and W. N. Johnson. 1991. Extent, cause, and timing of moose calf mortality in western interior Alaska. *Alces* 27:24–30.

Ozoga, J. J., and L. J. Verme. 1986. Relation of maternal age to fawn-rearing success in white-tailed deer. *J. Wildl. Mgmt.* 50:480–86.

Packard, J. M. 2003. Wolf behavior: reproductive, social, and intelligent. Pp. 35–65 in L. D. Mech and L Boitani, eds., *Wolves: behavior, ecology and conservation.* Chicago: University of Chicago Press.

Packer, C, and Ruttan L. 1988 The evolution of cooperative hunting. *Am. Nat.* 132:159–98.

Paquet, P. C. 1989. Behavioral ecology of sympatric wolves (*Canis lupus*) and coyotes (*C. latrans*) in Riding Mountain National Park, Manitoba. PhD thesis. University of Alberta, Edmonton.

Parker, G. R. 1977. Morphology, reproduction, diet, and behavior of the arctic hare (*Lepus arcticus monstrabilis*) on Axel Heiberg Island, Northwest Territories. *Can. Field Nat.* 91:8–18.

Parker, G. R., and S. Luttich. 1986. Characteristics of the wolf (*Canis lupus labradorius* Goldman) in northern Quebec and Labrador. *Arctic* 39:109–94.

Pasitchniak-Arts, M., M. E. Taylor, and L. D. Mech. 1988. Skeletal injuries in an adult arctic wolf. *Arct. and Alp. Res.* 20 (3): 360–65.

Patterson, B., and V. Power. 2002. Contributions of forage competition, harvest, and climate fluctuation to changes in population growth of northern white-tailed deer. *Oecologia* 130:62–71.

Peek, J. M. 1971. Moose habitat selection and relationships to forest management in northeastern Minnesota. PhD thesis. University of Minnesota, St. Paul.

———. 1982. Elk (*Cervus elaphus*). Pp. 851–61 in J. A. Chapman and G. A. Feldhamer, eds., *Wild mammals of North America: biology, management, and economics*. Baltimore, MD: John Hopkins University Press.

Peek, J. M., R. E. LeResche, and D. R. Stevens. 1974. Dynamics of moose aggregations in Alaska, Minnesota, and Montana. *J. Mammal.* 55:126–36.

Pembrey, M. 1996. Imprinting and transgenerational modulation of gene expression: human growth as a model. *Acta Geneticae Medicae et Genellologiae* 45:111–25.

Peters, R., and L. D. Mech. 1975. Behavior and intellectual adaptations of selected mammalian predators to the problem of hunting large animals. Pp. 279–300 in R. H. Tuttle, ed., *Socioecology and psychology of primates*. The Hague: Mouton.

Peterson, R. L. 1955. *North American moose*. Toronto: University of Toronto Press.

Peterson, R. O. 1977. *Wolf ecology and prey relationships on Isle Royale*. National Park Service Scientific Monograph Ser., no. 11. Washington, DC: Government Printing Office.

Peterson, R. O., and P. Ciucci. 2003. The wolf as a carnivore. Pp. 104–30 in L. D. Mech and L. Boitani, eds., *Wolves: behavior, ecology, and conservation*. Chicago: University of Chicago Press.

Peterson, R. O., and R. E. Page. 1983. Wolf-moose fluctuations at Isle Royale National Park, Michigan. *Acta Zool. Fenn.* 174:251–53.

Peterson, R. O., N. J. Thomas, J. M. Thurber, J. A. Vucetich, and T. A. Waite. 1998. Population limitation and the wolves of Isle Royale *J. Mammal.* 79:828–41.

Peterson, R. O., J. D. Woolington, and T. N. Bailey. 1984. *Wolves of the Kenai Peninsula, Alaska*. Wildlife Monograph no. 88. Bethesda, MD: Wildlife Society.

Pilot, M., W. Jedrzejewski, W. Branicki, V. E. Sidorovich, B. Jedrzejewska, K. Stachura, and S. M. Funk. 2006. Ecological factors influence population genetic structure of European grey wolves. *Mol. Ecol.* 15:4533–4553. doi: 10.1111/j.1365–294X.2006.03110.x.

Pimlott, D. H., J. A. Shannon, and C. B. Kolenosky. 1969. *The ecology of the timber wolf in Algonquin Park*. Research Report (Wildlife) 87. [Toronto]: Ontario Department of Lands and Forest, Research Branch.

Post, E., R. O. Peterson, N. C. Stenseth, and B. E. McLaren. 1999. Ecosystem consequences for wolf behavioural response to climate. *Nature* 401:905–7.

Post, E., and N. C. Stenseth. 1998. Large-scale climate fluctuation and population dynamics of moose and white-tailed deer. *J. Anim. Ecol.* 67:537–43.

Potvin, F., and H. Jolicoeur. 1988. Wolf diet and prey selectivity during two periods for deer in Quebec: decline versus expansion. *Can. J. Zool.* 66:1274–79.

Proffitt, K. M., J. A. Cunningham, K. L. Hamlin, and R. A. Garrott. 2014. Bottom-up and top-down influences on pregnancy rates and recruitment of northern Yellowstone elk. *J. Wildl. Mgmt.* 78:1383–93.

Pruitt, W. O., Jr. 1960. Locomotor speeds of some large northern mammals. *J. Mammal.* 41:112.

———. 1965. A flight releaser in wolf-caribou relations. *J. Mammal.* 46:350–51.

Range, F., and Z. Viranyi. 2011. Development of gaze following abilities in wolves (*Canis lupus*). *PLoS ONE* 6 (2): e16888.

Rau, M. E., and F. R. Caron. 1979. Parasite-induced susceptibility of moose to hunting. *Can. J. Zool.* 57:2466–68.

Reed, J. E., W. B. Ballard, P. S. Gipson, B. T. Kelly, P. R. Krausman, M. C. Wallace, and D. B. Wester. 2006. Diets of free-ranging Mexican gray wolves in Arizona and New Mexico. *Wildl. Soc. Bull.* 34:1127–33.

Rehfeldt, G. E., N. L. Crookston, C. Saenz-Romero, and E. M. Campbell. 2012. North American vegetation model for land-use planning in a changing climate: a solution to large classification problems. *Ecol. Appl.* 22:119–41.

Reig, S. 1993. Nonaggressive encounter between a wolf pack and a wild boar. *Mammalia* 57:451–53.

Ripple, W. J., and R. L. Beschta. 2004. Wolves and the ecology of fear: can predation risk structure ecosystems? *BioSci.* 54:755–66.

Rogers, L. L., L. D. Mech, D. Dawson, J. M. Peek, and M. Korb. 1980. Deer distribution in relation to wolf pack territory edges. *J. Wildl. Mgmt.* 44:253–58.

Rutter, R. J., and D. H. Pimlott. 1968. *The world of the wolf*. Philadelphia: J. B. Lippincott Co.

Sand, H., C. Wikenros, P. Wabakken, and O. Liberg. 2006. Effects of hunting group size, snow depth and age on the success of wolves hunting moose. *Anim. Behav.* 72:781–89.

Sauer, P. R. 1984. Physical characteristics. Pp. 73–90 in L. K. Halls, ed., *White-tailed deer ecology and management*. Harrisburg, PA: Stackpole Books.

Scheel, D, and C. Packer. 1991. Group hunting behavior of lions—a search for cooperation. *Anim. Behav.* 41:697–709.

Schaefer, J. A. 2008. Boreal forest caribou. Pp. 223–39 in M. Hummel and J. C. Ray, eds., *Caribou and the north: a shared future*. Toronto: Dundurn Press.

Schmidt, P. A., and L. D. Mech. 1997. Wolf pack size and food acquisition. *Am. Nat.* 150:513–17.

Schullery, P. 2004. *Searching for Yellowstone: ecology and wonder in the last wilderness*. Helena: Montana Historical Society Press.

Seal, U. S., M. E. Nelson, L. D. Mech, and R. L. Hoskinson. 1978. Metabolic indicators of habitat differences in four Minnesota deer populations. *J. Wildl. Mgmt.* 42:746–54.

Seton, E. T. 1927. Plains and wood bison. Pp. 369–717 in vol. 3 of *Lives of game animals*. Garden City, NY: Doubleday.

Severinghaus, C. W., and E. L. Cheatum. 1956. Life and times of the white-tailed deer. Pp. 57–186 in W. P. Taylor, ed., *The deer of North America*. Washington, DC: Stackpole Books.

Shapiro, B., A. J. Drummond, A. Rambaut, M. C. Wilson, P. E. Matheus, A. V. Sher, O. G. Pybus, et al. 2004. Rise and fall of the Beringian steppe bison. *Science* 306:1561–65.

Shaw J. H. 1995. How many bison originally populated western rangelands? *Rangelands* 17:148–50.

Sheldon, C. 1930. *The wilderness of Denali: explorations of a hunter-naturalist in northern Alaska.* New York: Charles Scribner's Sons.

Sih, A. S. 1980. Optimal behavior: can foragers balance two conflicting demands? *Science* 280:1041–43.

Smith, C. A. 1983. Responses of two groups of mountain goats, *Oreamnos americanus*, to the presence of a wolf, *Canis lupus*. *Can. Field Nat.* 97:110.

Smith, D. W., and E. E. Bangs. 2009. Reintroduction of wolves to Yellowstone National Park: history, values and ecosystem restoration. Pp. 92–125 in M. W. Hayward and M. J. Somers, eds., *Reintroduction of top-order predators*. Chicester: Wiley-Blackwell.

Smith, D. W., L. D. Mech, M. Meagher, W. E. Clark, R. Jaffe, M. K. Phillips, and J. A. Mack. 2000. Wolf-bison interactions in Yellowstone National Park. *J. Mammal.* 81:1128–35.

Smith, D. W., K. M. Murphy, and S. Monger. 2001. Killing of a bison (*Bison bison*) calf, by a wolf (*Canis lupus*), and four coyotes (*Canis latrans*), in Yellowstone National Park. *Can. Field Nat.*115 (2): 343–45.

Smith, D. W., D. R. Stahler, E. Albers, M. Metz, L. Williamson, N. Ehlers, K. Cassidy, et al. 2009. *Yellowstone wolf project: annual report, 2008.* YCR-2009-03. Yellowstone National Park, WY: National Park Service, Yellowstone Center for Resources.

Smith, T. G. 1980. Hunting, kill, and utilization of a caribou by a single gray wolf. *Can. Field Nat.* 94 (2): 175–77.

Smith, T. G., and I. Stirling. 1975. The breeding habitat of the ringed seal (*Phoca hispida*): the birth lair and associated structures. *Can. J. Zool.* 53:1297–1305.

Sokolov, V. E., A. S. Severtsov, and A. V. Shubkina. 1990. Modelling of the selective behavior of the predator towards prey: the use of borzois for catching saigas. [Translation from Russian.] *Zool. Zh.* 69:117–25.

Soper, J. D. 1941. History, range, and home life of the northern bison. *Ecol. Monogr.* 11:347–412.

Stanwell-Fletcher, J. F., and Stanwell-Fletcher, T. C. 1942. Three years in the wolves' wilderness. *Nat. Hist.* 49 (3): 136–47.

Steinberg, H. 1977. Behavioral responses of the white-tailed deer to olfactory stimulation using predator scents. Master's thesis. Pennsylvania State University, University Park.

Stenlund, M. H. 1955. A field study of the timber wolf (*Canis lupus*) on the Superior National Forest, Minnesota. *Minn. Dept. Cons. Tech. Bull.* 4:155.

Stephens, P. W., and R. O. Peterson. 1984. Wolf-avoidance strategies of moose. *Holarctic Ecol.* 7:239–44.

Stephenson, T. R., and V. Van Ballenberghe. 1995a. Defense of one twin calf against wolves, *Canis lupus*, by a female moose, *Alces alces*. *Can. Field Nat.* 109:251–53.

———. 1995b. Wolf, *Canis lupus*, predation on dusky Canada geese, *Branta canadensis occidentalis*. *Can. Field Nat.* 109:253–55.

Stringham, S. F. 1974. Mother-infant relations in moose. *Nat. Can.* 101:325–69.

Sumanik, R. S. 1987. Wolf ecology in the Kluane Region, Yukon Territory. Master's thesis. Michigan Technological University, Houghton.

Taylor, R. J., and P. J. Pekins. 1991. Territory boundary avoidance as a stabilizing factor in wolf-deer interactions. *Theor. Pop. Biol.* 39:115128.

Tener, J. S. 1954a. *A preliminary study of the musk-oxen of Fosheim Peninsula, Ellesmere Island, N.W.T.* Wildlife Management Bulletin, ser. 1, no. 9. Ottawa: Canada Dept. of Northern Affairs and National Resources, National Parks Branch, Canadian Wildlife Service.

———. 1954b. Three observations of predators attacking prey. No. 2. Polar wolf attacking arctic hares. *Can. Field Nat.* 68:180–82.

———. 1965. *Muskoxen in Canada.* Canada Wildlife Service Monograph Series, no. 2. Ottawa: Queens Printer.

Theberge, J. B., and M. T. Theberge. 2004. *The wolves of Algonquin Park: a 12 year ecological study.* Waterloo: Department of Geography, University of Waterloo.

Thompson, I. D., D. A. Welsh, and M. F. Vukelich. 1981. Traditional use of early winter concentration areas by moose in northeastern Ontario. *Alces* 17:1–14.

Thurber, J. M., and R. O. Peterson. 1993. Effects of population density and pack size on the foraging ecology of gray wolves. *J. Mammal.* 74:879–89.

Treisman, M. 1975. Predation and evolution of gregariousness. I. Models for concealment and evasion. *Anim. Behav.* 23:779–800.

Van Gelder, R. G. 1969. *Biology of mammals.* New York: Charles Scribner's Sons.

Verme, L. J. 1963. Effect of nutrition on growth of white-tailed deer fawns. *Trans. of North American Wildl. and Nat. Res. Conf.* 28:431–43.

Vucetich, J. A., M. Hebblewhite, D. W. Smith, and R. O. Peterson. 2011. Predicting prey population dynamics from kill rate and predation rate and predator-prey ratios in three wolf-ungulate systems. *J. Anim. Ecol.* 80:1236–45.

Vucetich, J. A., and R. O. Peterson. 2004. The influence of prey consumption and demographic stochasticity on population growth rate of Isle Royale wolves *Canis lupus*. *Oikos* 107:309–20.

Vucetich, J. A., D. W. Smith, and D. R. Stahler. 2005. Influence of harvest, climate and wolf predation on Yellowstone elk, 1961–2004. *Oikos* 111:259–70.

Wallace, L. L. 2004. *After the fires: the ecology of change in Yellowstone National Park.* New Haven, CT: Yale University Press.

Wang, X., and R. H. Tedford. 2008. *Dogs: their fossil relatives and evolutionary history.* New York: Columbia University Press.

Weaver, J. L. 1994. Ecology of wolf predation amidst high ungulate diversity in Jasper National Park, Alberta. PhD diss. University of Montana, Missoula.

Weaver, J. L., C. Arvidson, and P. Wood. 1992. Two wolves, *Canis lupus*, killed by a moose, *Alces alces*, in Jasper National Park, Alberta. *Can. Field Nat.* 106:126–27.

White, P. J., R. A. Garrott, K. L. Hamlin, R. C. Cook, J. G. Cook, and J. A. Cunningham. 2011. Body condition and pregnancy in northern Yellowstone elk: evidence for predation risk effects? *Ecol. Appl.* 21:3–8.

White, P. J., K. M. Proffitt, and T. O. Lemke. 2012. Changes in elk distribution and group sizes after wolf restoration. *Am. Midl. Nat.* 167:174–87.

White, P. J., K. M. Proffitt, L. D. Mech, S. B. Evans, J. A. Cunningham, and K. L. Hamlin. 2010. Migration of northern Yellowstone elk: implications for spatial structuring. *J. Mammal.* 91:827–37.

Wiebe, N., G. Samelius, R. T. Alisauskas, J. L. Bantle, C. Bergman, R. DeCarle, C. J. Hendrickson, A. Lusignan, K. J. Phipps, and J. Pitt. 2009. Foraging behaviours and diets of wolves in the Queen Maud Gulf Bird Sanctuary, Nunavut, Canada. *Arctic* 62:399–404.

Wikenros, C., H. Sand, P. Wabakken, O. Liberg, and H. C. Pedersen. 2009. Wolf predation on moose and roe deer: chase distances and outcome of encounters. *Acta Theriol.* 54:207–18.

Willis, C. M., S. M. Church, C. M. Guest, W. A. Cook, N. McCarthy, A. J. Bransbury, M. R. T. Church, and J. C. T. Church. 2004. Olfactory detection of human bladder cancer by dogs: proof of principle study. *Brit. Med. J.* 329:712. doi.org/10.1136/bmj.329.7468.712.

Wilton, M. L., and D. L Garner. 1991. Preliminary findings regarding elevation as a major factor in moose calving site selection in south central Ontario, Canada. *Alces* 27:111–17.

Wirsing, A. J. 2003. Predation mediated selection on prey morphology: a test using snowshoe hares. *Evol. Ecol. Res.* 5:315–27.

Wirsing, A. J., K. E. Cameron, and M. R. Heithaus. 2010. Spatial responses to predators vary with prey escape mode. *Anim. Behav.* 79:531–37.

Wishart, W. 1978. Bighorn sheep. Pp. 161–72 in J. L. Schmidt and D. L. Gilbert, eds., *Big game of North America: ecology and management.* Harrisburg, PA: Stackpole Books.

Young, S. P., and E. A. Goldman. 1944. *The wolves of North America.* Washington, DC: American Wildlife Institute.

Zamenhof, S., E. Van Marthens, and L. Grauel. 1971. DNA (cell number) in neonatal brain: second generation (F_2) alteration by maternal (F_0) dietary protein restriction. *Science* 172:850–51.

Author Index

Subject Index

The letter *f* following a page number denotes a figure, and the letter *t* denotes a table.